Lecture Notes in Networks and Systems

Volume 585

The series "Lecture Notes in Networks and Systems" publishes the latest developments in Networks and Systems—quickly, informally and with high quality. Original research reported in proceedings and post-proceedings represents the core of LNNS.

Volumes published in LNNS embrace all aspects and subfields of, as well as new challenges in, Networks and Systems.

The series contains proceedings and edited volumes in systems and networks, spanning the areas of Cyber-Physical Systems, Autonomous Systems, Sensor Networks, Control Systems, Energy Systems, Automotive Systems, Biological Systems, Vehicular Networking and Connected Vehicles, Aerospace Systems, Automation, Manufacturing, Smart Grids, Nonlinear Systems, Power Systems, Robotics, Social Systems, Economic Systems and other. Of particular value to both the contributors and the readership are the short publication timeframe and the world-wide distribution and exposure which enable both a wide and rapid dissemination of research output.

The series covers the theory, applications, and perspectives on the state of the art and future developments relevant to systems and networks, decision making, control, complex processes and related areas, as embedded in the fields of interdisciplinary and applied sciences, engineering, computer science, physics, economics, social, and life sciences, as well as the paradigms and methodologies behind them.

Indexed by SCOPUS, INSPEC, WTI Frankfurt eG, zbMATH, SCImago.

All books published in the series are submitted for consideration in Web of Science.

For proposals from Asia please contact Aninda Bose (aninda.bose@springer.com).

José Manuel Machado · Pablo Chamoso ·
Guillermo Hernández · Grzegorz Bocewicz ·
Roussanka Loukanova · Esteban Jove ·
Angel Martin del Rey · Michela Ricca
Editors

Distributed Computing and Artificial Intelligence, Special Sessions, 19th International Conference

Springer

Editors
José Manuel Machado
University of Minho
Braga, Portugal

Pablo Chamoso
University of Salamanca
Salamanca, Spain

Guillermo Hernández
University of Salamanca
Salamanca, Spain

Grzegorz Bocewicz
Koszalin University of Technology
Koszalin, Poland

Roussanka Loukanova
Stockholm University
Stockholm, Sweden

Bulgarian Academy of Sciences
Institute of Mathematics and Informatics
Sofia, Bulgaria

Esteban Jove
University of Coruña
La Coruña, Spain

Angel Martin del Rey
Departamento de Matemática Aplicada
University of Salamanca
Salamanca, Spain

Michela Ricca
University of Calabria
Arcavacata di Rende, Cosenza, Italy

ISSN 2367-3370 ISSN 2367-3389 (electronic)
Lecture Notes in Networks and Systems
ISBN 978-3-031-23209-1 ISBN 978-3-031-23210-7 (eBook)
https://doi.org/10.1007/978-3-031-23210-7

This Springer imprint is published by the registered company Springer Nature Switzerland AG
The registered company address is: Gewerbestrasse 11, 6330 Cham, Switzerland

Preface

Research on Intelligent Distributed Systems has matured during the last decade, and many effective applications are deployed now. Nowadays, technologies such as Internet of Things (IoT), Industrial Internet of Things (IIoT), Big Data, Blockchain, and distributed computing, in general, are changing constantly as a result of the large research and technical effort being undertaken in both universities and businesses. Most computing systems from personal laptops to edge/fog/cloud computing systems are available for parallel and distributed computing. Distributed computing performs an increasingly important role in modern signal/data processing, information fusion, and electronics engineering (e.g., electronic commerce, mobile communications, and wireless devices). Particularly, applying artificial intelligence in distributed environments is becoming an element of high added value and economic potential.

The 19th International Symposium on Distributed Computing and Artificial Intelligence 2022 (DCAI 2022) is a major forum for presentation of development and applications of innovative techniques in closely related areas. The exchange of ideas between scientists and technicians from both academic and business areas is essential to facilitate the development of systems that meet the demands of today's society. The technology transfer in this field is still a challenge, and for that reason, this type of contributions is specially considered in this symposium. DCAI 2022 brings in discussions and publications on development of innovative techniques to complex problems. This year's technical program presents both high quality and diversity, with contributions in well-established and evolving areas of research. Specifically, 46 papers were submitted, by authors from 28 different countries (Angola, Argentina, Bahrain, Bangladesh, Brazil, Bulgaria, Burkina Faso, Canada, Chile, Colombia, Denmark, Ecuador, Finland, France, Germany, India, Italy, Japan, Lebanon, Nigeria, Poland, Portugal, Qatar, Russia, Spain, Turkey, UK, USA), representing a truly "wide area network" of research activity. Moreover, DCAI 2022 Special Sessions have been a very useful tool in order to complement the regular program with new or emerging topics of particular interest to the participating community. The technical program of the Special Sessions of DCAI 2022 has selected 22 papers (12 full papers). As in past editions of DCAI, there will be special issues in highly ranked journals such as Electronics, Systems Journal, International Journal of Interactive Multimedia and

Artificial Intelligence, Smart Cities Journal and Advances in Distributed Computing and Artificial Intelligence Journal. These special issues will cover extended versions of the most highly regarded works, including from the Special Sessions of DCAI, which emphasize specialized, multi-disciplinary and transversal aspects. This year, DCAI 2022 has especially encouraged and welcomed contributions on: AI-driven methods for Multimodal Networks and Processes Modeling (AIMPM), Computational Linguistics, Information, Reasoning, and AI 2022 (CLIRAI), Intelligent Systems Applications (ISA), Mathematical Techniques in Artificial Intelligence and Machine Learning (MaTe-AI&ML), and New Perspectives and Solutions in Cultural Heritage (TECTONIC). Moreover, Doctoral Consortium Session tries to provide a framework as part of which students can present their ongoing research work and meet other students and researchers and obtain feedback on lines of research for the future.

We would like to thank all the contributing authors, the members of the Program Committee, the sponsors (IBM, Indra, Dipartimento di Ingegneria e Scienze dell'Informazione e Matematica dell'Università degli Studi dell'Aquila, Armundia Group, Whitehall Reply, T.C. Technologies And Comunication S.R.L., LCL Industria Grafica, AIR Institute, AEPIA, APPIA), and the Organizing Committee of the Universities of L'Aquila and Salamanca for their hard and highly valuable work. We are especially grateful for the funding supporting by project "XAI—XAI—Sistemas Inteligentes Auto Explicativos creados con Módulos de Mezcla de Expertos", ID SA082P20, financed by Junta Castilla y León, Consejería de Educación, and FEDER funds. And finally, we are grateful and value Program Committee members for their hard work, which has been essential for the success of DCAI 2022.

July 2022

José Manuel Machado
Pablo Chamoso
Guillermo Hernández
Grzegorz Bocewicz
Roussanka Loukanova
Esteban Jove
Angel Martin del Rey
Michela Ricca

Organization

Honorary Chairman

Sigeru Omatu	Hiroshima University, Japan

Advisory Board

Yuncheng Dong	Sichuan University, China
Francisco Herrera	University of Granada, Spain
Kenji Matsui	Osaka Institute of Technology, Japan
Tan Yigitcanlar	Queensland University of Technology, Australia
Tiancheng Li	Northwestern Polytechnical University, China

Program Committee Chairs

Rashid Mehmood	King Abdulaziz University, Saudi Arabia
Pawel Sitek	Kielce University of Technology, Poland

Organizing Committee Chairs

Sara Rodríguez	University of Salamanca, Spain
Serafino Cicerone	University of L'Aquila, Italy
Pablo Chamoso	University of Salamanca, Spain
Guillermo Hernández	University of Salamanca, Spain
José Manuel Machado	University of Minho, Portugal

Organizing Committee

Juan M. Corchado Rodríguez	University of Salamanca and AIR Institute, Spain
Fernando De la Prieta	University of Salamanca, Spain
Sara Rodríguez González	University of Salamanca, Spain
Javier Prieto Tejedor	University of Salamanca and AIR Institute, Spain
Pablo Chamoso Santos	University of Salamanca, Spain
Liliana Durón	University of Salamanca, Spain
Belén Pérez Lancho	University of Salamanca, Spain
Ana Belén Gil González	University of Salamanca, Spain
Ana De Luis Reboredo	University of Salamanca, Spain
Angélica González Arrieta	University of Salamanca, Spain
Emilio S. Corchado Rodríguez	University of Salamanca, Spain
Alfonso González Briones	University of Salamanca, Spain
Yeray Mezquita Martín	University of Salamanca, Spain
Beatriz Bellido	University of Salamanca, Spain
María Alonso	University of Salamanca, Spain
Sergio Marquez	University of Salamanca, Spain
Marta Plaza Hernández	University of Salamanca, Spain
Guillermo Hernández González	AIR Institute, Spain
Ricardo S. Alonso Rincón	University of Salamanca, Spain
Raúl López	University of Salamanca, Spain
Sergio Alonso	University of Salamanca, Spain
Andrea Gil	University of Salamanca, Spain
Javier Parra	University of Salamanca, Spain

Local Organizing Committee

Pierpaolo Vittorini (Co-chair)	University of L'Aquila, Italy
Tania Di Mascio (Co-chair)	University of L'Aquila, Italy
Federica Caruso	University of L'Aquila, Italy
Anna Maria Angelone	University of L'Aquila, Italy

DCAI 2022 Sponsors

Contents

Special Session on AI-driven Methods for Multimodal Networks and Processes Modeling (AIMPM'22)

The special session entitled AI-driven methods for Multimodal Networks and Processes Modeling (AIMPM 2022) is a forum that will share ideas, projects, researches results, models, experiences, applications, etc., associated with artificial intelligence solutions for different multimodal networks born problems (arising in transportation, telecommunication, manufacturing, and other kinds of logistic systems). The session held in L'Aquila (Italy) as the part of the 19th International Symposium Distributed Computing and Artificial Intelligence 2022.

Recently, a number of researchers involved in research on analysis and synthesis of multimodal networks devote their efforts to modeling different, real-life systems. The generic approaches based on the AI methods, highly developed in recent years, allow to integrate and synchronize different modes from different areas concerning: the transportation processes synchronization with concurrent manufacturing and cash ones or traffic flow congestion management in wireless mesh and ad hoc networks as well as an integration of different transportations networks (buses, rails, subway) with logistic processes of different character and nature (e.g., describing the overcrowded streams of people attending the mass sport and/or music performance events in the context of available holiday or daily traffic services routine). Due to the abovementioned reasons, the aim of the workshop is to provide a platform for discussion about the new solutions (regarding models, methods, knowledge representations, etc.) that might be applied in that domain. There is a number of emerging issues with big potential for methods of artificial intelligence (evolutionary algorithms, artificial neural networks, constraint programming, constraint logic programming, data-driven programming, answer set programming, hybrid methods—AI/OR-Operation Research, fuzzy sets) like multimodal processes management, modeling and planning production flow, production planning and scheduling, stochastic models in planning and controlling, simulation of discrete manufacturing system, supply chain management, mesh-like data network control, multimodal social networks, intelligent transport and passenger and vehicle routing, security of multimodal systems, network knowledge modeling, intelligent web mining and applications, business multimodal processes, and projects planning.

Organizing Committee

Chairs

Paweł Sitek, Kielce University of Technology, Poland
Grzegorz Bocewicz, Koszalin University of Technology, Poland
Izabela E. Nielsen, Aalborg University, Denmark

Co-chairs

Peter Nielsen, Aalborg University, Denmark
Zbigniew Banaszak, Koszalin University of Technology, Poland
Robert Wójcik, Wrocław University of Technology, Poland
Jarosław Wikarek, , Kielce University of Technology, Poland
Arkadiusz Gola, Lublin University of Technology, Poland
Mukund Nilakantan Janardhanan, University of Leicester, UK

Multimodal Network Based Graphs of Primitives Storage Concept for Web Mining CBIR

Tomasz Michno(✉) and Roman Stanislaw Deniziak

Kielce University of Technology, al. Tysiąclecia Państwa Polskiego 7, 25-314 Kielce, Poland
{t.michno,deniziak}@tu.kielce.pl

Abstract. Nowadays multimedia databases are becoming more and more popular. Because there are many images uploaded to the Internet there is a need of efficient querying and storing them. Additionally, a method of efficient web mining method also should be researched. The proper querying and storing data in a multimedia database is very crucial. In this paper a new concept of using Multimodal Network for storing and organizing graph of primitives for web mining Content Based Image Retrieval has been proposed. The graphs of primitives has been developed in our previous works in order to provide a new CBIR method. The usage of Multimodal Network would be beneficial for storing different types of nodes (e.g. nodes used only for organizing the hierarchy, nodes storing graph of primitives of objects or nodes storing image files). Additionally, the paper describes also modified K-Mean clustering method for gathering together in clusters similar graphs of primitives. Because of limited time, only preliminary experiments has been performed.

Keywords: Query by approximate shapes · Multimodal networks · Content based image retrieval

1 Introduction

Nowadays multimedia databases are becoming more and more popular. Because there are many images uploaded to the Internet there is a need of efficient querying and storing them. Additionally, a method of efficient web mining method also should be researched [4].

The proper querying and storing data in a multimedia database is very crucial, thus in our previous works we prepared some initial researches for the Content Based Image Retrieval algorithm in [5], crawling and retrieving images in [3] and the graphs of primitives used for objects description (which are defined as a modified graph with additional properties [5]) comparisons in [4].

Through the years researchers developed many graph storage approaches. There are methods which uses relational databases with SQL [13,16] and NoSQL

J. M. Machado et al. (Eds.): DCAI 2022, LNNS 585, pp. 3–12, 2023.
https://doi.org/10.1007/978-3-031-23210-7_1

approaches [1,6]. Graphs may be also stored using hierarchic structure, for example [12]. An example approach which is dedicated to distributed graph storage systems was described in [20]. Moreover there are some transactional approaches, for example [21] and dedicated fault tolerancy and scalability [11]. For more efficient storage there are also used approaches like for example reinforcement learning [19] or graph neural network [18]. Another approach may be using Multimodal Network in order to organize different graphs.

Multimodal Networks have become more and more popular in researches in recent years. They are designed on a basis of graphs an hypergraphs in order to add capabilities for representing relationships which are present in biological networks or databases [7]. The definition of a Multimodal Network extends the classical graph 2-tuple (Vertices and Edges) to 3-tuple (Vertices, Modal hyperedges and Modes) [7]. A modal hyperedge is an edge which has a specified type and a type is defined by a mode [8]. This specific type of relationship may be very useful in designing organization of a database of graph of primitives, which will be described in more details in next sections. Multimodal Networks and their variations are mainly used in biology [7], transport [9], vehicle routing problems [15,17], supply chains [14] and batch production flows [2].

The main motivation for this paper is to prepare initial concept for the graphs of primitives storage in Multimodal Network which is able to represent different relations between nodes used for organizing graphs, nodes storing graphs and nodes connected with storing image files. In the future research this concept will be extended into distributed approach. Additionally, for gathering similar graphs of primitives in similar part of the network, K-Mean based clustering method is used.

This paper has the following structure: in the Sect. 2 the idea of graphs of primitives is described. The Sect. 3 describes the modified K-Mean clustering method and the Sect. 4 presents the concept of using a Multimodal Network for storing graphs of primitives. In the Sect. 5 preliminary experimental results are shown. The last section summarizes the idea and paper.

2 Graphs of Primitives

During previous works on Query by Approximate Shapes algorithm [5] different methods of sketch representation has been considered and as a result a representation using predefined set of shapes has been chosen. The set of predefined shapes can be defined as [5]:

$$T = \{L, A, PL, PG, Al, AL\} \tag{1}$$

where: O - a line, A - an arc, PL - a poly line, PG - a polygon, AL - a chain of connected arches, AG - a looped chain of connected arches.

For each of this predefined shapes we can define its property (e.g. for line its slope can be used or for an arc its angle), creating a primitive which can be defined as:

$$p_i = (t_i, a_i) : t_i \in T, a_i \subset A \tag{2}$$

where: t_i - primitive type, $t_i \in T$, A - a set of primitives attributes, defined as $A = \{a_{i1}, a_{i2}, a_{i3}, ...a_{in} : a_{i1}...a_{in} \in [0,1], n \in N_+\}$.

In order to store mutual relations between primitives they are stored in a form of a modified graph, where each node is dedicated to one primitive and edges are used to represent connections between primitives. Additionally, the locations of primitives are stored using the geographical directions set K (containing following values: N, S, W, E and their combinations). The graph of primitives can be then defined as:

$$G_i = (V_i, E_i), \tag{3}$$

$$V_i = \{p_{ik} : p_{ik} \in P\}, \tag{4}$$

$$E_i = \{(p_{ia}, p_{ib}, k) : p_{ia} \in V_i \wedge p_{ib} \in V_i \wedge p_{ia} \neq p_{ib} \wedge k \in K\} \tag{5}$$

where: V_i - a set of nodes, E_i - a set of edges between nodes with information about their mutual location.

3 Gathering Similar Graphs of Primitives Using Modified K-Mean Clustering

One of the most common Unsupervised Machine Learning usage is finding clusters in a data which groups similar objects together. This task can be performed even for data which is with unknown classes but has similar properties.

The K-Mean clustering divides data into so called clusters which are groups of data points which have small variance with the centroid of the cluster. Mostly squared Euclidean distances is used to partition data. The K-Mean clustering in a modified version may be also used to gather similar graphs of primitives in clusters. In order to do that, two operations has to be defined: measuring the distance between two graphs of primitives and computing the mean of graphs of primitives.

Measuring the distance between two graphs of primitives may be defined as:

$$dist(G_1, G_2) = 1 - sim \tag{6}$$

$$sim(G_1, G_2) \in < 0, 1 > \tag{7}$$

where *sim* is a similarity of two graphs, which may be computed using different algorithms described e.g. in [4]. This implies that when two graphs of primitives are the same, the *sim* value is equal to 1 and the *dist* is equal to 0. When two graphs of primitives are completely different, the *sim* value is equal to 0 and the *dist* is equal to 1. For all other situations, the values $\in (0, 1)$.

Another operation which has to be defined for using K-Mean clustering for graphs of primitives is computing the mean graph of primitives, which can be defined as computing the average of each graph in cluster intersections:

$$G_m = avg(G_1 \cap G_2, G_2 \cap G_3, G_1 \cap G_3, ..., G_{n-1} \cap G_n) \tag{8}$$

where: G_1, G_2, ... , G_n - graphs of primitives for which the average graph has to be computed.

The intersection of two graphs of primitives G_1 and G_2 can be defined as finding a set of vertices and edges which has the highest *sim* and lowest *dist* coefficients with both G_1 and G_2 graphs:

$$G_1 \cap G_2 = \begin{aligned} \min_{G_m,\xi} \quad & dist(G_1 \cap G_2, G_1) + dist(G_1 \cap G_2, G_2) \\ \text{s.t.} \quad & dist(G_1, G_2) \leq \xi \\ & dist(G_1 \cap G_2, G_1) \leq \xi \\ & dist(G_1 \cap G_2, G_2) \leq \xi \\ & \xi \geq 0 \end{aligned} \tag{9}$$

where: ξ - the maximum distance between two graphs of primitives which are considered as similar.

In order to compute average graph of primitives, there should be also defined the *avg* operation as follows:

1. Create empty average graph of primitives G_{avg}.
2. For each mean graph G_{m_i} given by parameter to $avg(G_{m_1}, G_{m_2}, ..., G_{m_i}, ..., G_{m_n})$:
   (a) for each pair of vertices $v_{m_{il}}$ and $v_{m_{ik}}$ and edge between them $e_{m_{ilk}}$ of G_{m_i}:
      (i) check if at least a half of other mean graphs contains similar vertexes and edges as $v_{m_{il}}$ and $v_{m_{ik}}$ and $e_{m_{ilk}}$ (compute *sim* using Eq. 7 and check if obtained value is $\geq (1 - \xi)$)
      (ii) if the condition is met, add $v_{m_{il}}$ and $v_{m_{ik}}$ and $e_{m_{ilk}}$ to G_{avg} if they are not present in the average graph.
3. return G_{avg}.

An example of K-Mean clustering for graphs of primitives is presented in the Fig. 1. The step 1 is dedicated to choosing randomly from existing graphs means - initial centroids for clusters. Next in the Step 2 the distances using Eq. 6 is computed and depending on results all graphs are assigned to each cluster. In the Step 3 new centroids are computed for each cluster: for cluster 1 - new graph of primitives is created as a mean of all graphs from the cluster, using Eq. 8, for cluster 2 - because there is only one graph in this cluster, the mean remains the same. In the Step 4 distances are computed using new centroids and all graphs are assigned according to the results. Next, in the Step 5 new centroids are computed (because there are no new mean graphs of primitives for centroids, all centroids remain without changes). After that, in the Step 6 distances are computed and assignments to clusters are made. Because there are no changes, the algorithm converges and finishes.

4 Multimodal Network for Query by Approximate Shapes

Multimodal Networks are very powerful tools for representing mutual dependencies between nodes which are hard to cover using classical graph or tree representation.

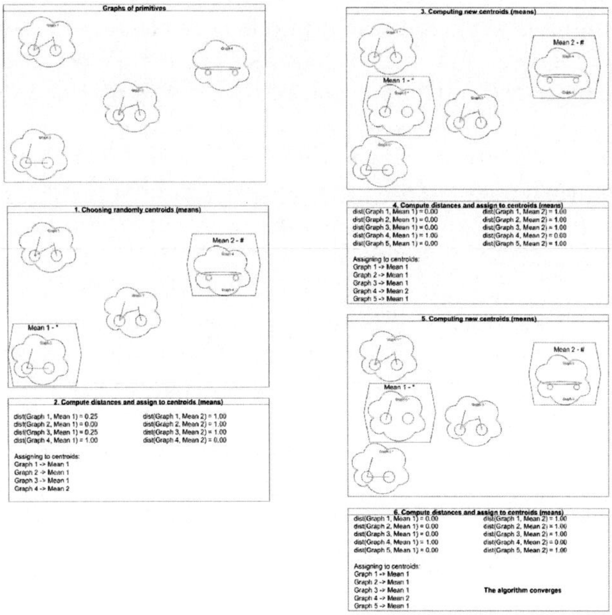

Fig. 1. Example of K-Mean clustering for graphs of primitives.

Table 1. The results of K-Mean clustering.

Iteration	Graphs in cluster 1	Graphs in cluster 2	Graphs in cluster 3	Graphs in cluster 4
0	–	–	–	–
1	1, 5	2, 3, 4	6, 7	8, 9, 10
2	1, 5	2, 4, 6	3, 7	8, 9, 10
3	1, 5	2, 4, 6	3, 7	8, 9, 10

In this work only preliminary approach was considered, but in the future more advanced properties of Multimodal Networks will be considered. Additionally distributed processing for Network will be taken into account.

The architecture of the Multimodal Network was designed in order to preserve all elements which are needed for Query by Approximate Shapes algorithm: information about objects graphs of primitives, their connection with image files and also mean graphs of primitives in order to gather similar graphs of primitives together. Because of that, there could be defined following types of nodes: entry node - a node used at first when traversing the network nodes, mean graph of primitives - a node which is dedicated to gathering common parts of other graphs stored in the network, cannot store connections to image files, object graph of primitives - a node which stores graph of primitives of objects present in image files, image file node - a node which is dedicated to storing one image file (the file physically may be stored in different ways e.g. in the memory or as a link to file on a disc), graph of primitives - is a special set of nodes which has the structure described in Chap. 2 and is a part of Multimodal Network.

In order to cover different types of nodes, there should be also defined different types of edges: from entry node to mean graphs of primitives nodes, from mean graphs of primitives nodes to objects graphs of primitives nodes, from means and objects graph of primitives nodes to graph of primitives, from objects graphs of primitives nodes to image file nodes.

An example of proposed Multimodal Network architecture is presented in the Fig. 2. The Node number 0 is an entry node, thus it is the first node which is entered after start traversing the network. From Entry Node there are present edges to Mean graph nodes (nodes: 1, 2 and 3). Mean graph nodes have edges to object graph of primitives nodes (nodes: 4, 5, 6, 7, 8, 9). Both mean graph nodes and object graph nodes have connections to adequate nodes in graph of primitives nodes set. Object graph of primitives nodes may be connected to one or more image file nodes (e.g. like in nodes 4, 10 and 11, where the same object appears in two different image files). Moreover, there could be also similar case when one image file node is connected to different object nodes (e.g. like in nodes 12, 5 and 6, where the image contains two different objects).

The major disadvantages of proposed approach are: long new images/graphs addition time caused i.e. by inner structure reorganizing (which may be a problem for some applications) and problem with storing huge amount of data. Both drawbacks will be investigated in future research, which may include using some distributed approaches both for speed increase and storage efficiency.

5 Preliminary Experimental Results

Thus, the Multimodal Network usage for Query by Approximate Shapes is an initial concept, only preliminary experiments were performed for testing the usability of K-Mean based clustering algorithm. The tests were performed for a set of 10 different graphs of primitives which are shown if the Fig. 3 and 4 clusters. The test results are presented in the Table 1. As an initial centroids for the 0 iteration following graphs has been chosen: Graph no. 1, Graph no. 3, Graph no. 7, Graph no. 9 (all centroids after each iteration are shown in the Fig. 4). For each

Table 2. Preliminary real life example results.

Object	Query by approximate shapes		Regions		Kato et al. [10]	
	Precision	Recall	Precision	Recall	Precision	Recall
Car 1 (Fiat 500)	0.89	0.33	0.53	0.75	0.58	0.83
Car 2 (Mercedes Benz)	0.79	0.73	0.51	0.5	0.47	0.7
Bike	0.93	0.37	0.23	0.42	0.4	0.5
Scooter	0.86	0.40	0.75	0.4	0.25	0.5

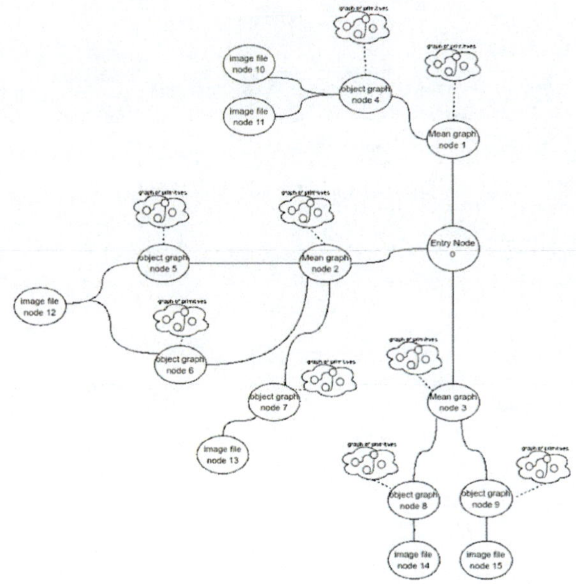

Fig. 2. Example multimodal network for graphs of primitives.

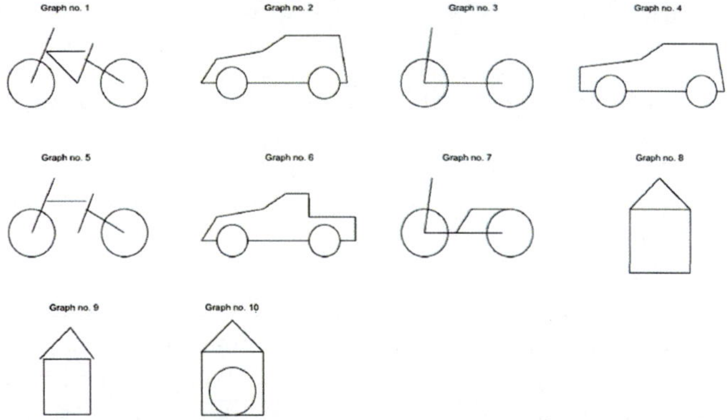

Fig. 3. Graphs of primitives used for experiments.

algorithm iteration distances have been computed, graphs have been assigned to clusters and new centroids have been chosen. The algorithm converges after 3 iterations. Additionally the initial results from prototype application has been provided in the Table 2, but due to limited implementation time, they lack K-Mean clustering.

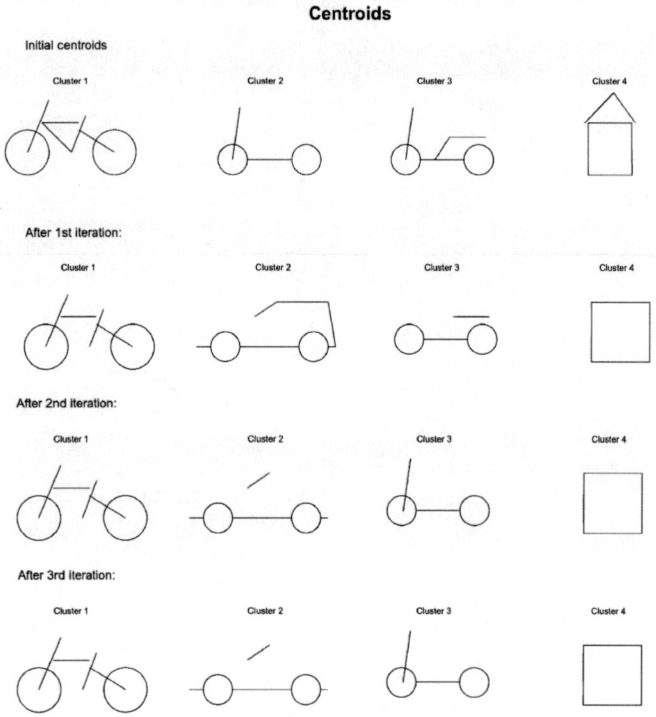

Fig. 4. Mean graphs - centroids in different iterations for the test graph of primitives set.

6 Summary

In this paper a new concept of using Multimodal Network for storing and organizing graph of primitives for web mining Content Based Image Retrieval has been proposed. The usage of such a structure would be beneficial for storing different types of nodes (e.g. nodes used only for organizing the hierarchy, nodes storing graph of primitives of objects or nodes storing image files). Additionally, the paper describes also modified K-Mean clustering method for gathering together in clusters similar graphs of primitives. Because of limited time, only preliminary experiments has been performed.

In the future research development there should be added hierarchical K-Mean based clustering algorithm in order to more efficiently gather similar graphs of primitives in near parts of Multimodal Network. Additionally, some distributed approaches for the Network will be proposed. Another area which should be investigated is the usage of Multimodal Edges in graph of primitives.

References

1. Agrawal, S., Patel, A.: A study on graph storage database of NOSQL. Int. J. Soft Comput. Artif. Intell. Appl. **5**, 33–39 (2016). https://doi.org/10.5121/ijscai.2016.5104
2. Bocewicz, G., Nielsen, I., Smutnicki, C., Banaszak, Z.: Towards the leveling of multi-product batch production flows. A multimodal networks perspective. IFAC-PapersOnLine **51**(11), 1434–1441 (2018). 16th IFAC Symposium on Information Control Problems in Manufacturing INCOM 2018. https://doi.org/10.1016/j.ifacol.2018.08.313, www.sciencedirect.com/science/article/pii/S240589631831437X
3. Deniziak, R.S., Michno, T.: World wide web CBIR searching using query by approximate shapes. In: Rodríguez, S., Prieto, J., Faria, P., Kłos, S., Fernández, A., Mazuelas, S., Jiménez-López, M.D., Moreno, M.N., Navarro, E.M. (eds.) 15th International Conference on Distributed Computing and Artificial Intelligence, Special Sessions, pp. 87–95. Springer International Publishing, Cham (2019)
4. Deniziak, R.S., Michno, T.: Graph of primitives matching problem in the world wide web CBIR searching using query by approximate shapes. In: Herrera-Viedma, E., Vale, Z., Nielsen, P., Martin Del Rey, A., Casado Vara, R. (eds.) 16th International Conference on Distributed Computing and Artificial Intelligence, Special Sessions, pp. 77–84. Springer International Publishing, Cham (2020)
5. Deniziak, S., Michno, T.: Query by approximate shapes image retrieval with improved object sketch extraction algorithm. In: 2018 Federated Conference on Computer Science and Information Systems (FedCSIS), pp. 555–559 (2018)
6. Fosić, I., Šolić, K.: Graph database approach for data storing, presentation and manipulation. In: 2019 42nd International Convention on Information and Communication Technology, Electronics and Microelectronics (MIPRO), pp. 1548–1552 (2019). https://doi.org/10.23919/MIPRO.2019.8756793
7. Heath, L.S., Sioson, A.A.: Multimodal networks: structure and operations. IEEE/ACM Trans. Comput. Biol. Bioinform. **6**(2), 321–332 (2009). https://doi.org/10.1109/TCBB.2007.70243
8. Heath, L.S., Sioson, A.A.: Semantics of multimodal network models. IEEE/ACM Trans. Comput. Biol. Bioinform. **6**(2), 271–280 (2009). https://doi.org/10.1109/TCBB.2007.70242
9. Huang, H., Bucher, D., Kissling, J., Weibel, R., Raubal, M.: Multimodal route planning with public transport and carpooling. IEEE Trans. Intell. Transp. Syst. **20**(9), 3513–3525 (2019). https://doi.org/10.1109/TITS.2018.2876570
10. Kato, T., Kurita, T., Otsu, N., Hirata, K.: A sketch retrieval method for full color image database-query by visual example. In: 11th IAPR International Conference on Pattern Recognition, Vol. I. Conference A: Computer Vision and Applications, pp. 530–533 (Aug 1992)
11. Malewicz, G., Austern, M.H., Bik, A.J., Dehnert, J.C., Horn, I., Leiser, N., Czajkowski, G.: Pregel: a system for large-scale graph processing. In: Proceedings of the 2010 ACM SIGMOD International Conference on Management of Data, SIGMOD'10, pp. 135–146. Association for Computing Machinery, New York, NY, USA (2010). https://doi.org/10.1145/1807167.1807184
12. Pandey, P., Wheatman, B., Xu, H., Buluc, A.: Terrace: a hierarchical graph container for skewed dynamic graphs, pp. 1372–1385. Association for Computing Machinery, New York, NY, USA (2021). https://doi.org/10.1145/3448016.3457313

13. Schmid, M.: An approach to efficiently storing property graphs in relational databases. In: Grundlagen von Datenbanken (2019)
14. Sitek, P., Wikarek, J.: Cost optimization of supply chain with multimodal transport. In: 2012 Federated Conference on Computer Science and Information Systems (FedCSIS), pp. 1111–1118 (2012)
15. Sitek, P., Wikarek, J., Rutczyńska-Wdowiak, K., Bocewicz, G., Banaszak, Z.: Optimization of capacitated vehicle routing problem with alternative delivery, pick-up and time windows: a modified hybrid approach. Neurocomputing **423**, 670–678 (2021)
16. Sun, W., Fokoue, A., Srinivas, K., Kementsietsidis, A., Hu, G., Xie, G.: Sqlgraph: an efficient relational-based property graph store. In: Proceedings of the 2015 ACM SIGMOD International Conference on Management of Data, SIGMOD'15, pp. 1887–1901. Association for Computing Machinery, New York, NY, USA (2015). https://doi.org/10.1145/2723372.2723732
17. Thibbotuwawa, A., Nielsen, P., Zbigniew, B., Bocewicz, G.: Energy consumption in unmanned aerial vehicles: a review of energy consumption models and their relation to the UAV routing. In: Świątek, J., Borzemski, L., Wilimowska, Z. (eds.) Information Systems Architecture and Technology: Proceedings of 39th International Conference on Information Systems Architecture and Technology-ISAT 2018, pp. 173–184. Springer International Publishing, Cham (2019)
18. Wang, M., Yu, L., Zheng, D., Gan, Q., Gai, Y., Ye, Z., Li, M., Zhou, J., Huang, Q., Ma, C., Huang, Z., Guo, Q., Zhang, H., Lin, H., Zhao, J., Li, J., Smola, A.J., Zhang, Z.: Deep graph library: towards efficient and scalable deep learning on graphs. CoRR (2019). arxiv.org/abs/1909.01315
19. Yuan, G., Lu, J., Zhang, S., Yan, Z.: Storing multi-model data in RDBMSs based on reinforcement learning, pp. 3608–3611. Association for Computing Machinery, New York, NY, USA (2021). https://doi.org/10.1145/3459637.3482191
20. Zhang, W., Chen, Y., Dai, D.: Akin: a streaming graph partitioning algorithm for distributed graph storage systems. In: 2018 18th IEEE/ACM International Symposium on Cluster, Cloud and Grid Computing (CCGRID), pp. 183–192 (2018). https://doi.org/10.1109/CCGRID.2018.00033
21. Zhu, X., Feng, G., Serafini, M., Ma, X., Yu, J., Xie, L., Aboulnaga, A., Chen, W.: Livegraph: a transactional graph storage system with purely sequential adjacency list scans. CoRR (2019). arxiv.org/abs/1910.05773

Race Condition Error Detection in a Program Executed on a Device with Limited Memory Resources

Rafał Wojszczyk(✉), Damian Giebas, and Grzegorz Bocewicz

Koszalin University of Technology, Sniadeckich 2, 75-453 Koszalin, Poland
rafal.wojszczyk@tu.koszalin.pl

Abstract. Wearable devices are a prime example of how electronics are invading people's everyday lives. This is forcing electronics manufacturers to use hardware that is cheap to manufacture, allows for long battery life, and is also computationally efficient. Combining these requirements is very challenging. A compromise can lie in the use of low-cost microcontrollers programmed in native C-type language. Additionally, using at least dual-core microcontrollers will allow the implementation of software that runs efficiently and responsively. Unfortunately, software using multiple cores is prone to errors related to resource conflicts. This paper presents an example of detecting one such error that occurred in a dual-core microcontroller.

Keywords: Multithreading programming · Race condition · Microcontrollers

1 Introduction

The history of technical development includes several events that are considered milestones, e.g., the invention of the steam engine. In the case of computers and computing, development is usually driven by large companies such as Microsoft, Apple, and Google. Occasionally, it happens that one person, or more precisely, that person's ideas get noticed and spread to the market. This situation has been changed by the idea of crowdfunding, which is the support of society for ideas (start-ups) proposed by ordinary people. It consists in the fact that one person (or a group of people) wants to offer a product or service, however, it requires funding for most of the costs, including development, production and dissemination. Ideas implemented in start-ups should not only fit into the expectations of the community, but also be credible. Credibility is especially important when making promises for electronic devices. To this end, start-ups build prototypes of devices that consist of fairly cheap and widely available components, such as popular Arduino-compatible microcontrollers. These kinds of solutions are also used in manufacturing devices and science [1].

J. M. Machado et al. (Eds.): DCAI 2022, LNNS 585, pp. 13–20, 2023.
https://doi.org/10.1007/978-3-031-23210-7_2

Building your own device dedicated to the consumer market requires a lot of skills from the creator. It should include programming skills, knowledge of electronics basics, communication protocols and many other issues related to computer science and electronics. These are significant requirements that must be met by the aforementioned one person or a small team. In addition, devices aimed at such a market often require fast and responsive operation, integration with smartphone applications and data storage in the cloud. At the same time, it is important to keep manufacturing costs as low as possible. In a way, this leads to a conflict of interest. High-performance devices (e.g. those using NVIDIA Tegra chips, Jetson) provide ease of implementation in modern programming languages (e.g. C#, Lua, Python), but at the same time are expensive to manufacture and often have high power consumption, which is very important in mobile devices. On the other hand, low-cost devices (e.g. ESP32 with 2×240 MHz clock and 520 KB SRAM) are more difficult to program (they usually use C or assembler language), but do not share the drawbacks of more expensive solutions. In this context, a certain compromise is to use low-cost devices with multithreaded programming capabilities. Delegation of selected operations to separate threads allows for efficiency and expansiveness effects even in low-cost devices, e.g., a smartband that simultaneously measures human vital signs and displays a notification about a new message.

The purpose of this paper is to discuss the process of detecting errors that cause resource conflicts (e.g., harmful race condition, deadlock, atomicity violation, and order violation) in devices built on popular dual-core microcontrollers. Such devices are built by enthusiasts, student research clubs, or in the aforementioned start-ups. Building such devices requires the said knowledge from various fields and various skills. This paper presents a solution that can be used in the software testing stage. The developed RD-L method can be applied to applications written in C language using pthread library and devices (microcontrollers) with limited memory.

The next Sect. 2 presents selected methods enabling analysis of multithreaded applications. Section 3 discusses the program in which the race condition class error occurred. Then, in the Sect. 4, an author's proposal for detecting errors of this type is presented. The last chapter contains the summary.

2 Literature Review

The elements of multithreading in computer programs should be tested in one process with the other types of tests. Unfortunately, this is often not possible because of the long time that must elapse for an error to occur in a multithreaded application. The long time is due to the fact that many specific conditions must be met for an error to occur, e.g., inter-thread interactions occurring, threads running simultaneously by the operating system scheduler. This leads to a situation where manual and unit tests are basically useless.

Methods of testing multithreaded applications known from the literature are often based on dynamic software analysis ([2,3] or [4]). It consists in monitoring the program operation, which is performed by a supervisor - an additional program. Methods of this type cannot be used for applications and devices mentioned in the introduction of the paper (technical limitations related to the amount of memory and computing power will not allow it).

Methods outside the testing domain consist in preventing the occurrence of multithreading errors [5,6]. Methods of this type have similar limitations as methods based on dynamic software analysis. Additionally, they require interference with the compiler, which unnecessarily increases code maintenance costs.

The last group of methods that can be used to detect multithreading errors are methods based on static software analysis. Methods of this type analyze the source code and therefore are not limited by the amount of memory of the target device. Source code analysis is usually limited to a particular programming language, which means that methods dedicated to Java and C# (e.g. [7] or [8]) cannot be used for C language. On the other hand, methods that can be used for C language only detect selected multithreading errors (e.g. [9] for race condition and deadlock, [10] for race condition) or in case of detecting at least several types of multithreading errors prove to be ineffective [5].

In the context presented, it can be seen that there is still a lack of solutions that can be used for microcontrollers programmed in C (or a C twin) with simultaneous detection of four classes of errors: race condition, deadlock, order violation, and atomicity violation.

3 Error Occurrence

Among the four errors that lead to resource conflicts of the types: race condition, deadlock, order violation, and atomicity violation, the first two are the most common. The occurrence of an error discussed next concerns race condition.

The occurrence of race condition can be caused by many different phenomena, depending on what data the ongoing threads are competing for. In the optimistic (least harmful) case, some of the temporary data will be damaged, which will be manifested, for example, by small artifacts in the video. The worst case scenario may even lead to the machine being taken over by unauthorized persons [11]. For Arduino devices using the pthread library, race condition can occur just like in any other application using this library. Software for Arduino devices is structured a bit differently than multi-threaded applications written in C language for example for Linux. This is due to the fact that the code written for Arduino devices is written in C++. Every programmer, who knows both languages, knows that "small differences" between them may cause "big problems". In any case, writing code on Arduino that looks like C the programmer actually uses C++ language mechanisms. However, practice shows that programs written on Arduino often do not use elements characteristic for C++ language and resemble programs following structural paradigm. It is worth mentioning, that before popularization of Arduino platform, the most popular language for programming ATMEL AVR microcontrollers was C and avr-gcc compiler. Therefore, if a programmer does not explicitly use C++ language elements in his code (i.e. does not define classes, namespaces etc.) and limits himself/herself to paradigms supported by C language (which is a common practice in low-level device support), they can use the RD-L method. It should be noted here that using classes e.g. "Serial.println();" looks the same as using a structure that contains a pointer to a function. So, if all the classes developed for Arduino have an interface holding this scheme, then it is possible to use the RD-L method.

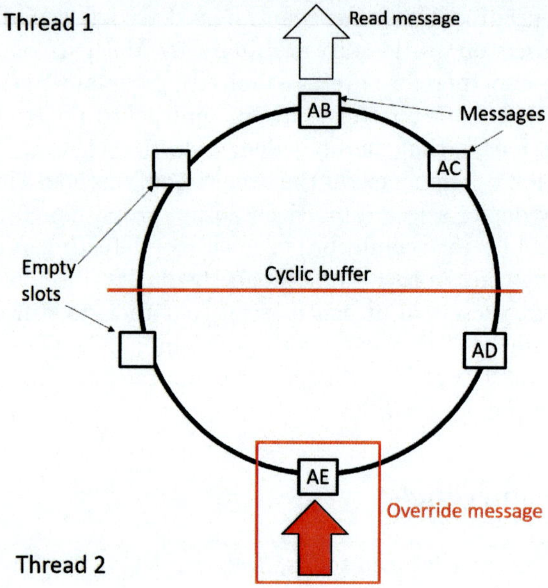

Fig. 1. Presentation of message overwriting in a cyclic buffer.

With the above assumptions met, an ESP32 family microcontroller was programmed using Arduino with a bug that leads to race condition. The error occurred in a student project for the construction of a small self-driving chassis. The chassis can be used to build simple self-driving devices to which appropriate heads are mounted, e.g. for 3d scanning, a camera to record a moving object. The chassis is equipped with ultrasonic distance sensors that detect obstacles along the path of the vehicle. The vehicle has been built using several microcontrollers. The master microcontroller (e.g. other ESP32, Raspberry PI, etc.) acts as the main controller, while additional ESP32 slave microcontrollers act as the executive systems (reading data from sensors, controlling motors, etc.). Two threads are implemented in the slave microcontroller, the first to communicate with the master microcontroller via UART, and the second used to handle connected devices. The first thread handles requests from the master microcontroller and reports interruptions when appropriate information is reported by the second thread. The second thread infinitely checks the state of the sensors and when it detects an anomaly (e.g. lack of ground under one of the wheels) it adds an appropriate message to the message buffer, see Fig. 1. The first thread has to receive such a message and, based on it, take actions which will not lead to undesirable behavior of the vehicle (e.g. overturning, hitting an obstacle). This mechanism, however was written incorrectly, i.e. in the second thread the operation placing data in the message buffer did not use mutexes. It means that this thread was adding messages in an uncontrolled way. Additionally, the thread did not check that the cyclic buffer being used was not full. This could lead to an error where the other thread could overwrite a previously posted message. This led to a situation where the first thread did not receive important messages (although this was the main intention when implementing multithreading), thus not reacting appropriately to events, and this eventually led to an accident.

The second thread implemented infinite sensor data reading, which is passed to the first thread. Unfortunately, the operation of the slave microcontroller turned out to be different from the scenario the developers expected.

4 Error Detection

Detecting the error described in the previous point comes down to locating the structure shown in Fig. 2. As mentioned before, the code in question is written in C++ language (but writing code according to C rules), where classes and method calls are treated as structures, whose components are pointers to functions. This simple assumption causes that no changes are required either in the solution code (apart from renaming the input function to main, which results from the limitation of the used tool called rdao detector), or in the RD-L method, or in the source code model of multithreaded applications.

The source code model of a multithreaded application contains the following structure [12]:

$$C_P = (T_P, U_P, R_P, O_P, Q_P, F_P, B_P) \tag{1}$$

where:

1. P is the program index,
2. T_P is the set of all threads,
3. U_P is a sequence of sets containing threads working in the same time frame,
4. R_P is a family of shared resources,
5. O_P is a set of all atomic operations of the program,
6. Q_P is the set of mutexes available in the program,
7. F_P is the set of edges,
8. B_P is sequence of sets; where every set contains pairs of operations related by forward relationship or backward relationship or symmetrical relation.

Figure 2 shows a visualization of the constructed model, where dashed vertical lines denote time intervals, circles denote operations, arrows denote transition relationships between operations, and a rectangle is a shared resource.

Analysis of the source code revealed the presence of a structure leading to a race condition at 200 ms. From the description of the error in the previous section it is known that this error was not limited only to the lack of mutexes, i.e., from the expert analysis it is known that the other thread also did not check if there is space for a new message in the cyclic buffer. Hence, in addition to making the changes necessary to eliminate race condition, changes had to be made to the message buffer handling. These changes were not trivial. Using cyclic buffer always involves a risk of overwriting existing messages. So the buffer had to be replaced with a FIFO queue, implemented as a one-way list. This approach allowed placing new messages at the end of the list without the risk of overwriting any of the messages already placed in the list. The only limitation in this case is the amount of device memory. If the device runs out of memory it becomes impossible to add a new message to the queue. Hundreds of messages were analyzed to solve this problem. As a result it was possible to group messages into a few different groups. Thanks to that instead of putting long messages with identical content in queue it was decided to use series of codes with 1 byte size. This approach results in memory savings because a 1 byte code carries the same information as a long, multi-byte message.

The error found was therefore very complex, i.e., apart from the error leading to race condition, it was necessary to change the data structure in which the messages were placed and to limit the message content in order to save memory. Unfortunately, it is impossible to automatically detect errors that involve the use of incorrect data structures. For this purpose, it was necessary to analyze source code by an expert.

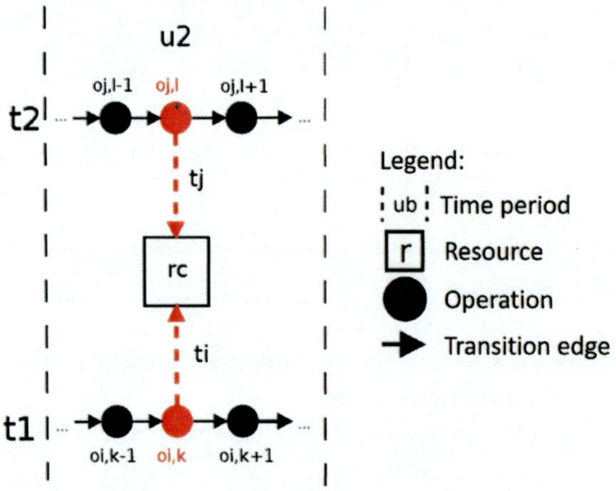

Fig. 2. Structure of an error leading to race condition, where $u_2 \in U_P, o_j, l, o_i, k \in O_P, t_i, t_j \in T_P, r_c \in R_P$.

5 Conclusion

This paper presents a case of detection of a race condition class error, which is one of the more common bugs found in multithreaded applications. The programmer's intention, which prompted him/her to use multithreading, was to ensure high reliability in operation (e.g. in a situation when one of the sensors will require a longer read time, which is the case with the 1-wire interface). Unfortunately, the presented case showed that an effect completely different from the assumed scenario was obtained, which emphasizes how important multithreading error detection is.

In the context of the goal of this paper, a race condition class error was successfully detected. This confirmed the usefulness of the developed RD-L method also when applied to a selected class of programs written in C++, what is a main contribution of this paper. The source code of the developed tool is available in the github repository (https://github.com/PKPhdDG/rdao_detector), so that everyone can use the tool for their own needs. The obtained results give reason to believe that the RD-L method can be applied to most programs written in C++ in the future, which is the main goal of further research. For this purpose, it is planned to develop algorithms for transforming the source code into the proposed model. The considered programming languages are primarily object-oriented languages, e.g. C ++, C# [13], and in the future other paradigms, e.g. CLP [14]. Adaptation to other languages may require an extension of the model, however, it will be done with backward compatibility. Another way of research is the adaptation of the model to areas other than programming, e.g. scheduling supply chains [15].

References

1. Wojszczyk, R., Giebas, D.: Repair of multithreaded errors in the control and measurement system. In: Distributed Computing and Artificial Intelligence. Vol. 2: Special Sessions 18th International Conference, DCAI 2021. Lecture Notes in Networks and Systems, vol. 332. Springer, Cham (2022). https://doi.org/10.1007/978-3-030-86887-1_4
2. Park, J., Choi, B., Jang, S.: Dynamic analysis method for concurrency bugs in multi-process/multi-thread environments. Int. J. Parallel Program. **48**(6), 1032–1060 (2020). https://doi.org/10.1007/s10766-020-00661-3
3. Lu, S., Tucek, J., Qin, F., Zhou, Y.: AVIO: detecting atomicity violations via access interleaving invariants. ACM SIGOPS Oper. Syst. Rev. **40**(5), 37–48 (2006). Association for Computing Machinery. https://doi.org/10.1145/1168917.116886
4. Park, S., Lu, S., Zhou, Y.: CTrigger: exposing atomicity violation bugs from their hiding places. In: Proceedings of the 14th International Conference on Architectural Support for Programming Languages and Operating Systems, vol. 44, no. 3, pp. 25–36. Association for Computing Machinery (2009). https://doi.org/10.1145/1508244.1508249
5. Yu, Z., Zuo, Y., Zhao, Y.: Convoider: a concurrency bug avoider based on transparent software transactional memory. Int. J. Parallel Program. **48**(1), 32–60 (2019). https://doi.org/10.1007/s10766-019-00642-1
6. Yu, W., Gao, F., Wang, L., Yu, T., Zhao, J., Li, X.: Automatic detection, validation and repair of race conditions in interrupt-driven embedded software. IEEE Trans. Softw. Eng. (2020). https://doi.org/10.1109/TSE.2020.2989171
7. Yi, J., Sadowski, C., Flanagan, C.: SideTrack: generalizing dynamic atomicity analysis. In: Proceedings of the 7th Workshop on Parallel and Distributed Sys-

tems: Testing, Analysis, and Debugging, pp. 1–10 (2009). https://doi.org/10.1145/1639622.1639630

8. Mathur, U., Viswanathan, M.: Atomicity checking in linear time using vector clocks. In: Proceedings of the Twenty-Fifth International Conference on Architectural Support for Programming Languages and Operating Systems, pp. 183–199. Association for Computing Machinery (2020). https://doi.org/10.1145/3373376.3378475
9. Engler, D., Ashcraft, K.: RacerX: effective, static detection of race conditions and deadlocks. ACM SIGOPS Oper. Syst. Rev. **37**(5), 237–252. Association for Computing Machinery (2003). https://doi.org/10.1145/1165389.945468
10. Qi, D., Gu, N., Su, J.: Detecting data race in network applications using static analysis. In: 2019 International Conference on Networking and Network Applications (NaNA), pp. 313–318. IEEE (2019). https://doi.org/10.1109/NaNA.2019.00061
11. Sekurak Homepage. https://sekurak.pl/atak-race-condition-przyklad-zastosowania-w-aplikacji-webowej/. Last accessed 28 Apr 2022
12. Giebas, D., Wojszczyk, R.: Detection of concurrency errors in multithreaded applications based on static source code analysis. IEEE Access (2021). https://doi.org/10.1109/ACCESS.2021.3073859
13. Wojszczyk, R.: The model and function of quality assessment of implementation of design patterns. Appl. Comput. Sci. **11**(3), 44–55 (2015)
14. Sitek, P., Wikarek, J.: A novel integrated approach to the modelling and solving of the two-echelon capacitated vehicle routing problem. Prod. Manuf. Res. **2**(1), 326–340 (2014). https://doi.org/10.1080/21693277.2014.910716
15. Chodorek, A., Chodorek, R.R., Sitek, P.: UAV-based and WebRTC-based open universal framework to monitor urban and industrial areas. Sensors **21**(12), Art. no. 4061 (2021). https://doi.org/10.3390/s21124061

The Use of Corporate Architecture in Planning and Automation of Production Processes

Zbigniew Juzoń(✉), Jarosław Wikarek, and Paweł Sitek

Kielce University of Technology, Al. Tysiąclecia P.P. 7, 25-314 Kielce, Poland
{zjuzon,j.wikarek,sitek}@tu.kielce.pl

Abstract. Production planning is a difficult and very complex issue; therefore, it is necessary to consider the enterprise as a whole, taking into account the relations between its elements. The paper presents the application of corporate architecture supported by constraint logic programming and mathematical programming to plan production processes as well as to build a corporate bus (service bus). Using the assumptions of corporate architecture and building a metamodel of corporate architecture based on good practices, it is much easier and more efficient to identify those processes that can be automated and robotized. The presented metamodel can be the basis for building detailed models for planning and automation of production processes.

Keywords: Corporate architecture · Production planning · Metamodel · Constraint logic programming · Mathematical programming

1 Introduction

In Industry 4.0, not only automation plays a very important role, but also full computerization and the use of various types of data collection, processing and exchange systems, which boils down to effective factory management and optimization of production processes [1]. It is necessary to take into account various uncertain factors (fluctuations in the value of global economic indicators, changes in the market environment, fluctuations in the value of revenues, costs, etc.). The occurrence of such situations means the need to develop new methods and systems supporting decision making, taking into account the uncertain factors in the system under consideration. It is necessary to consider the enterprise as a whole, taking into account the relations that occur between its components. Changing the approach to date allows the production system to identify, group production resources, define types of individual relationships and manage a set of these relationships. Sharing architectural knowledge between different organizations, or even within one organization (between its individual component units), is a big challenge. Architectural models are prepared in different languages (e.g., ArchiMate, UML, BPMN) [2–4] and with different levels of detail. As a result, there are difficulties with maintaining consistency when creating models of enterprise architecture (the larger and more complex the organization is, the greater the challenges in this area). The solution to this situation is the development of architectural metamodels, which are the basis

J. M. Machado et al. (Eds.): DCAI 2022, LNNS 585, pp. 21–32, 2023.
https://doi.org/10.1007/978-3-031-23210-7_3

for communication and exchange of architectural knowledge; the metamodel itself is defined as an explicit model for constructing domain models [5]. Below, an architecture metamodel (1) is proposed, which combines resources and processes with a set of relations in terms of corporate architecture:

$$\boldsymbol{M} = (\boldsymbol{R_s}, \boldsymbol{P}, \boldsymbol{R}) \tag{1}$$

where:

- $\boldsymbol{R_s}$ **resources** (*all material and non-material elements of the production process that are necessary to produce products, e.g., machines, raw materials, employees, tools, etc.*)
- $\boldsymbol{P}$ **processes** (*all phenomena and deliberately undertaken actions which result in the gradual occurrence of the desired changes in the subject of work subject to their influence*)
- $\boldsymbol{R}$ **relationships** (*all connections and interdependent that affect the manufacture or maintenance of products or services*)

This metamodel defines the semantics and how to use the concepts that appear on the detail models of planning, scheduling, resource allocation and other production processes. By using the created metamodel, it is possible to ensure a common understanding of the concepts by the stakeholders involved in the process of building corporate architecture.

2 Illustrative Example

To illustrate the discussed topic, an exemplary production process has been presented. The production facility manufactures products using 3 various industrial robots. The following designations for production resources were adopted (machine 1, machine 2, machine 3). An illustrative example shows a common problem of production planning and resource allocation, that is usually modelled and solved by using mathematical programming. Therefore, this example is ideally suited to show the relationship between a mathematical model and an architectural metamodel. The production volume of each manufactured product has been determined (Table 2).

In addition, the production of each product requires a certain amount of time on each machine. The production time is given in Table 1. A dash means that the production process of the product does not require a given machine.

The identification and elimination of bottlenecks in processes enables better management of production, resources and costs. This is especially important in distributed organizations - for a company that has several factories located in different countries, it may be important to build a coherent system of supply of raw materials and materials and to supervise the effectiveness of individual stages of production and distribution processes. Therefore, in this context, the questions below should be answered for the company to better respond to the market demand.

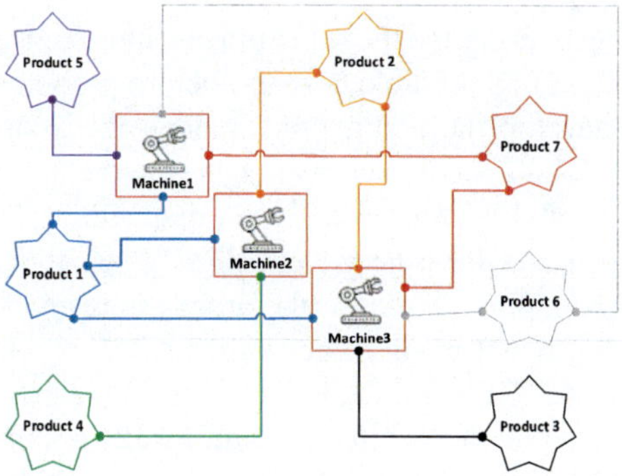

Fig. 1. Set of relationships (machine - product)

Table 1. Production process (data in the table are given in minutes)

Resources	Products						
	p_1	p_2	p_3	p_4	p_5	p_6	p_7
Machine 1	5	7	–	–	3	2	5
Machine 2	1	2	–	3	–	6	–
Machine 3	2	–	8	–	–	–	6

Table 2. Sales plan (the quantity of the product was given in pieces)

Month	Products						
	p_1	p_2	p_3	p_4	p_5	p_6	p_7
January	30	–	–	–	20	10	–
February	–	10	20	–	30	10	–
March		40	–	–	–	20	–
April	40	–	20	400	–	10	30
May	20	–	10	20	–	–	20
June	–	–	20	10	–	–	–

Q1 **How to plan production and maintenance to meet the sales plan, keep maintenance and storage costs as low as possible?**

Q2 **What parameters of the Service Level Agreement (SLA) can it guarantee and introduce to new product supply contracts?**

3 Architecture Model

In order to answer questions (Q1/Q2), implementation is necessary, corporate architecture and an architecture model was built to optimize the production system, which allowed for the grouping of tasks and the selection of technological processes for automation, within which it would be possible to manufacture products using various types of industrial robots. To develop the architecture model, the assumptions resulting from the TOGAF [6] methodology were used, which define the architectural framework for describing the corporate architecture of a given organization. In particular, at the technology layer level, use cases have been defined as a set of scenarios related to each other by a common goal. When building the architecture model, it was assumed that the production of each product would require the use of specific production resources and business requirements. Corporate data bus (Fig. 2) means a data repository, it is a place where data obtained using architecture and results obtained by solving decision and optimization problems are stored. Information gathered at the corporate bus level can be reused at the organizational level, at the design stage of future SLA. SLA are created to document the obligations that need to be fulfilled to customers.

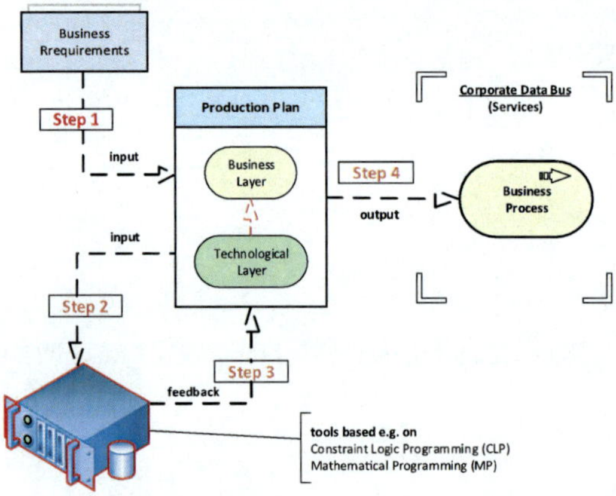

Fig. 2. The elements of corporate architecture according to TOGAF (*step by step*)

In the proposed solution concept, the process is as follows **Step 1:** All the above-mentioned components of the architecture metamodel (Fig. 2), identification of production resource types, identification of relationship types, identification of types of production processes for the needs, can then be considered as an argument for investment programming. **Step 2:** The process of building an architectural model taking into account the approach, the service and the layer is not an automatic process and during its construction, first of all, you should build a set of good practices for modelling corporate architecture that has proven successful in various projects [7]. We start the process of creating an architecture with building a model of the motivational layer and leads to the construction of corporate services (Fig. 2) in order to get the answer to question Q2. **Step 3:** To answer question Q1, a detailed planning and resource allocation model was formulated (Sect. 4) for an exemplary production process (Fig. 1) **Step 4:** It comes down to building a corporate data bus.

4 Mathematical Model of Production Planning and Resource Allocation

To answer question Q1: *How to plan production and maintenance to meet the sales plan, keep maintenance and storage costs as low as possible?* - it is necessary to build a mathematical model. The mathematical model was formulated for optimization and problem solving using Mixed-integer linear programming (MILP). Table 4 defines the elements of the mathematical model.

Fulfilment of customer orders

$$U_{p,t-1} + X_{p,t} + Y_{p,t} = a_{p,t} + U_{p,t} \forall p \in P, t \in T - \{t_o\} s_p + X_{p,t} + Y_{p,t} \\ = a_{p,t} + U_{p,t} \forall p \in P, t = t_o \quad (2)$$

The load of machines only within the permitted limits

$$\sum_{p \in P} \left(b_{p,r} \cdot X_{p,t}\right) = o_{r,t} \cdot K_{r,t} \forall r \in R, t \in T \quad (3)$$

Performing scheduled maintenances

$$\sum_{t \in T} \left(1 - K_{r,t}\right) = h_r - F_r \forall r \in R \quad (4)$$

Only the permitted number of maintenances during the period

$$\sum_{r \in R} \left(1 - K_{r,t}\right) = h_m \forall t \in T \quad (5)$$

The quantity of the product in the warehouse does not exceed the allowed value

$$U_{p,t} = v_p \forall p \in P, t \in T \quad (6)$$

Total storage capacity not exceeded

$$\sum_{p \in P} \left(U_{p,t} - z_p\right) \leq v_m \forall t \in T \quad (7)$$

Calculation of storage costs

$$Cost = \sum_{p \in P} \sum_{t \in T} \left(U_{p,t} - g_p\right) \quad (8)$$

Determination of the value of L_1

$$L_1 = \sum_{r \in R} F_r \quad (9)$$

How many total products have not been made for all orders

$$L_2 = \sum_{p \in P} \sum_{t \in T} Y_{p,t} \quad (10)$$

Table 3. Defining elements of the mathematical model

Architecture metamodel elements	Symbol	Model elements	Description
R_s	P	Sets & indexes	Product $p \in P$
	R		Machine $r \in R$
	T		Time period $t \in T$, t_o - initial period, t_k - end period
P	$X_{p,t}$	Decision variables	Production volume of the product p in the period t
	$Y_{p,t}$		What order quantity for the product p in period t we are not able to fulfill
	$K_{r,t}$		If the machine is not to be maintenance in period t $K_{r,t} = 1$ else $K_{r,t} = 0$
	F_r		If the overhaul of the machine r cannot be performed, then $F_r = 1$ else $F_r = 0$
R	$U_{p,t}$	Determined values	Stock of product p at the end of period t
	Cost		Product storage cost
	L_1		How many scheduled maintenances have not been carried out
	L_2		How many products in total have not been comp. for all orders
	2	Constraints	Fulfilment of customer orders
	3		The load of machines only within the permitted limits
	4		Performing scheduled maintenances
	5		Only the permitted number of maintenances during the period
	6		The quantity of the product in the warehouse does not exceed the allowed value
	7		Total storage capacity not exceeded
	8		Calculation of storage costs

(*continued*)

Table 3. (*continued*)

Architecture metamodel elements	Symbol	Model elements	Description
	9		Determination of the value of L_1
	10		How many products in total have not been made for all orders
	11		Binary and integrity

Binary and integrity

$$
\begin{aligned}
X_{p,t} &\in N \forall p \in P, t \in T \\
Y_{p,t} &\in N \forall p \in P, t \in T \\
U_{p,t} &\in N \forall p \in P, t \in T \\
K_{r,t} &\in \{0, 1\} \forall r \in R, t \in T \\
F_r &\in \{0, 1\} \forall r \in R
\end{aligned}
\tag{11}
$$

5 Numerical Experiment

Formalization of the mathematical model (Sect. 4) shows that the number of equations and decision variables is so large that looking for solution by hand is not practical, due to the time-consuming nature and the possibility of mistakes. To solve this problem, it is recommended to use software tools. A proprietary hybrid approach that integrates Constraint Logic Programming and Mathematical Programming was proposed for the implementation and solution of the modelled problem [10, 11]. The ECLiPSe CLP [8] was used to generate the model, which was then solved with the Gurobi solver [9]. The *Experiment_1* and *Experiment_2* was conducted on a workstation having the following specification: Processor: Intel(R) Core (TM) i7-10700K CPU @ 3.80 GHz 3.79 GHz; RAM: 16 GB; Windows 11; Processor x64. In order to verify the use of the mathematical model, the data from Tables 1, 2, 4, 5 and 6 that was taken into account on *Experiment_1* and *Experiment_2*. The duration of the computational experiments, taking into account the data from Tables 1, 2, 4, 5 and 6 ranged from 2 to 5 s (Table 3).

The results presented in Tables 7 and 8 allow us to obtain the best possible production and renovation plan, so that the total cost of product storage is as low as possible. The conducted experiments and the results obtained in Tables 7 and 8 also provided other key information that the implementation of certain contracts and maintenances is impossible. For example, for *Experiment_1*, the use of the model showed that the entire production plan would be successful, but renovation for machines 1 and 3 would not be successful. However, for *Experiment_2*, it will not be possible to fully fulfil orders for the product p1, p4 and p6 in period 4. Similarly, renovation for machines 1 and 3 will not be successful. It should be emphasized that this is important information for the business side and must be taken into account at the level of future contracts, which will define the so-called

Table 4. Data for *Experiment_1* and *Experiment_2*

Description	Value
Planning horizon - T	6
Number of products - P	7
Number of machines - R	3
Total storage capacity -$\mathbf{v_m}$	1000
Max. number of renovations -$\mathbf{h_m}$	1
The maximum stock of the product in the warehouse -$\mathbf{v_p}$	10 for $p_1 \ldots p_2$
Product storage volume -$\mathbf{z_p}$	2 for $p_1,\ p_2$ 3 for $p_3 \ldots p_7$
Initial stock of the product -$\mathbf{s_p}$	1 for $p_1 \ldots p_7$
Product storage cost **p** per time unit -$\mathbf{g_p}$	0.5 for $p_1 \ldots p_7$
Is there any planned overhaul of the machine? – $\mathbf{h_r}$	1 for $r_1 .. r_3$
How much time does it take to process the product p on the machine $\mathbf{b_{p,r}} \forall \mathbf{p} \in \mathbf{P}, \mathbf{t} \in \mathbf{T}$	0 for $b_{p_3,r_1}, b_{p_4,r_1}, b_{p_3,r_2}, b_{p_5,r_2}, b_{p_7,r_2}, b_{p_2,r_3}, b_{p_4,r_3}, b_{p_5,r_3}, b_{p_6,r_3}$ 1 for b_{p_1,r_2} 2 for $b_{p_6,r_1}, b_{p_2,r_2}, b_{p_1,r_3}$ 3 for b_{p_5,r_1}, b_{p_4,r_2} 5 for b_{p_1,r_1}, b_{p_7,r_1} 6 for b_{p_6,r_1}, b_{p_7,r_3} 7 for b_{p_2,r_1} 8 for b_{p_3,r_3}
Machine production capacity r in the period $\mathbf{o_{r,t}} \forall \mathbf{r} \in \mathbf{R}, \mathbf{t} \in \mathbf{T}$	1000 for $r_1 \ldots r_3$

Table 5. Additional data for *Experiment_1*

Description	Values selected for the needs of the *Experiment_1*
Penalty for failure to perform maintenances -$\mathbf{K_{r_1}}$	100000
Penalty for failure to fulfill orders $\mathbf{K_{r_2}}$	1000

SLA parameters. The use of the architecture model in conjunction with a detailed model can used to solve the planning problem or optimize production and maintenances. In the proposed solution concept, the metamodel of the corporate architecture (Fig. 2) provides a complete set of information, the so-called input data for a detailed model (mathematical

Table 6. Additional data for *Experiment_2*

Description	Values selected for the needs of the *Experiment_2*
Penalty for failure to perform maintenances -$\mathbf{K_{r_1}}$	1000000
Penalty for failure to fulfill orders $\mathbf{K_{r_2}}$	10

model) with which we can plan or optimize the production system, taking into account all constraints and conditions resulting from the metamodel of the corporate architecture.

6 Conclusion

Based on the assumptions of the corporate architecture resulting from the TOGAF standard, it is possible to build a model of the motivation layer to provide important input for a future mathematical model. At the stage of analysing the model of the motivation layer, he determines, at a high level of abstraction, the result that should be obtained to achieve the goals. The results obtained during the implementation of *Experiment_1* (Table 7) and *Experiment_2* (Table 8) showed that the modification of the previously selected parameters at the stage of the architecture metamodel contained in (Table 5) or (Table 6) allows for the determination of SLA parameters for the needs of new contracts for the delivery of products and building an optimal production and maintenances plan. Outcomes are high-level, business-oriented outcomes that result from the organization's capabilities. In general, to be able to verify the initially adopted theoretical assumptions, a detailed mathematical model should be developed (similar to the presented in Sect. 4), which should be solved using a tool based on artificial intelligence methods and/or Mathematical Programming (MP), and Constraint Logic Programming (CLP) [10–12] to support the decision-making process. The tools based on CLP and/or MP are ideal for solving problems of less complexity. For complex problems Artificial Neural Networks (ANN) and/or Genetic Algorithms (GA) can be used [13]. Therefore, at the stage of future research, it is worth experimenting in the field of teaching Artificial Neural Networks (ANN) supported by CLP and MP.

Appendix A. Checking the Model for the Data from the Illustrative Example

See Tables 7 and 8.

Table 7. Production plan - results of the *Experiment_1*

[p1] period = 1 init_stock = 0 order = 30 prod. = 30 unrealized order = 0 final stock = 0
[p1] period = 3 init_stock = 0 order = 0 prod. = 40 unrealized order = 0 final stock = 40
[p1] period = 4 init_stock = 40 order = 40 prod. = 0 unrealized order = 0 final stock = 0
[p1] period = 5 init_stock = 0 order = 20 prod. = 20 unrealized order = 0 final stock = 0
[p2] period = 2 init_stock = 0 order = 10 prod. = 10 unrealized order = 0 final stock = 0
[p2] period = 3 init_stock = 0 order = 40 prod. = 40 unrealized order = 0 final stock = 0
[p3] period = 2 init_stock = 0 order = 20 prod. = 20 unrealized order = 0 final stock = 0
[p3] period = 4 init_stock = 0 order = 20 prod. = 20 unrealized order = 0 final stock = 0
[p3] period = 5 init_stock = 0 order = 10 prod. = 10 unrealized order = 0 final stock = 0
[p3] period = 6 init_stock = 0 order = 20 prod. = 20 unrealized order = 0 final stock = 0
[p4] period = 3 init_stock = 0 order = 0 prod. = 100 unrealized order = 0 final stock = 100
[p4] period = 4 init_stock = 100 order = 400 prod. = 300 unrealized order = 0 final stock = 0
[p4] period = 5 init_stock = 0 order = 20 prod. = 20 unrealized order = 0 final stock = 0
[p4] period = 6 init_stock = 0 order = 10 prod. = 10 unrealized order = 0 final stock = 0
[p5] period = 1 init_stock = 0 order = 20 prod. = 20 unrealized order = 0 final stock = 0
[p5] period = 2 init_stock = 0 order = 30 prod. = 30 unrealized order = 0 final stock = 0
[p6] period = 1 init_stock = 0 order = 10 prod. = 10 unrealized order = 0 final stock = 0
[p6] period = 2 init_stock = 0 order = 10 prod. = 10 unrealized order = 0 final stock = 0
[p6] period = 3 init_stock = 0 order = 20 prod. = 20 unrealized order = 0 final stock = 0
[p6] period = 4 init_stock = 0 order = 10 prod. = 10 unrealized order = 0 final stock = 0
[p7] period = 4 init_stock = 0 order = 30 prod. = 30 unrealized order = 0 final stock = 0
[p7] period = 5 init_stock = 0 order = 20 prod. = 20 unrealized order = 0 final stock = 0
Renovations not made: for machines = 1 and for machines = 3 ≫ **Storage cost = 18000**

Table 8. Production plan - results of the *Experiment_2*

[p1] period = 3 init_stock = 0 order = 0 prod. = 5 unrealized order = 0 final stock = 5
[p1] period = 4 init_stock = 5 order = 40 prod. = 0 unrealized order = 35 final stock = 0
[p1] period = 5 init_stock = 0 order = 20 prod. = 20 unrealized order = 0 final stock = 0
[p2] period = 2 init_stock = 0 order = 10 prod. = 10 unrealized order = 0 final stock = 0
[p2] period = 3 init_stock = 0 order = 40 prod. = 40 unrealized order = 0 final stock = 0
[p3] period = 2 init_stock = 0 order = 20 prod. = 20 unrealized order = 0 final stock = 0
[p3] period = 4 init_stock = 0 order = 20 prod. = 20 unrealized order = 0 final stock = 0
[p3] period = 5 init_stock = 0 order = 10 prod. = 10 unrealized order = 0 final stock = 0
[p3] period = 6 init_stock = 0 order = 20 prod. = 20 unrealized order = 0 final stock = 0
[p4] period = 4 init_stock = 0 order = 400 prod. = 326 unrealized order = 74 final stock = 0
[p4] period = 5 init_stock = 0 order = 20 prod. = 20 unrealized order = 0 final stock = 0
[p4] period = 6 init_stock = 0 order = 10 prod. = 10 unrealized order = 0 final stock = 0
[p5] period = 1 init_stock = 0 order = 20 prod. = 20 unrealized order = 0 final stock = 0
[p5] period = 2 init_stock = 0 order = 30 prod. = 30 unrealized order = 0 final stock = 0
[p6] period = 1 init_stock = 0 order = 10 prod. = 10 unrealized order = 0 final stock = 0
[p6] period = 2 init_stock = 0 order = 10 prod. = 10 unrealized order = 0 final stock = 0
[p6] period = 3 init_stock = 0 order = 20 prod. = 20 unrealized order = 0 final stock = 0
[p6] period = 4 init_stock = 0 order = 10 prod. = 0 unrealized order = 10 final stock = 0
[p7] period = 4 init_stock = 0 order = 30 prod. = 30 unrealized order = 0 final stock = 0
[p7] period = 5 init_stock = 0 order = 20 prod. = 20 unrealized order = 0 final stock = 0
Unrealized order: period = 4 ⋙ [p1] = 35 and [p4] = 74 and [p6] = 10
Renovations not made: for machines = 1 and for machines = 3 ⋙ **Storage cost = 1000**

References

1. Scrimieri, D., Afazov, S.M., Ratchev, S.M.: Design of a self-learning multi-agent framework for the adaptation of modular production systems. Int. J. Adv. Manuf. Technol. **115**(5–6), 1745–1761 (2021). https://doi.org/10.1007/s00170-021-07028-z
2. Stiehl, V.: Process-Driven Applications with BPMN (2014). https://doi.org/10.1007/978-3-319-07218-0, ISBN 978-3-319-07218-0
3. Opekunova, L.A., Opekunov, A.N., Kamardin, I.N., Nikitina, N.V.: Modeling enterprise architecture using language ArchiMate (2019). https://doi.org/10.1007/978-3-030-27015-5_61, ISBN 978-3-030-27014-8
4. Rumpe, B.: Modeling with UML (2016). https://doi.org/10.1007/978-3-319-33933-7, ISBN 978-3-319-81635-7
5. Sobczak, A.: Models and Metamodels in Corporate Architecture (2008)
6. The Open Group Architecture Framework (TOGAF) Standard. https://pubs.opengroup.org/architecture/togaf9-doc/arch/index.html. Last accessed 2018
7. Greefhorst, D., Proper, E.: Architecture Principles—The Cornerstones of Enterprise Architecture (2011). https://doi.org/10.1007/978-3-642-20279-7, ISSN 1867-8920

8. ECLiPSe Constraint Logic Programming System. www.eclipseclp.org. Last accessed 2022
9. Gurobi. https://www.gurobi.com. Last accessed 2022
10. Sitek, P., Wikarek, J.: A multi-level approach to ubiquitous modeling and solving constraints in combinatorial optimization problems in production and distribution. Appl. Intell. **48**(5), 1344–1367 (2017). https://doi.org/10.1007/s10489-017-1107-9
11. Sitek, P., Wikarek, J.: A hybrid programming framework for modeling and solving constraint satisfaction and optimization problems. Sci. Program. **2016**, ID 5102616, 13 pages (2016). https://doi.org/10.1155/2016/5102616
12. Thibbotuwawa, A., Bocewicz, G., Nielsen, P., Banaszak, Z.: Planning deliveries with UAV routing under weather forecast and energy consumption constraints. In: 9th IFAC Conference on Manufacturing Modelling, Management and Control MIM 2019, IFAC-PapersOnLine, vol. 52(13), pp. 820–825 (2019). https://doi.org/10.1016/j.ifacol.2019.11.231
13. Świć, A., Wołos, D., Gola, A., Kłosowski, G.: The use of neural networks and genetic algorithms to control low rigidity shafts machining. Sensors **20**, 4683 (2020). https://doi.org/10.3390/s20174683

Special Session on Computational Linguistics, Information, Reasoning, and AI 2022 (CompLingInfoReasAI'22)

Computational and technological developments that incorporate natural language and reasoning methods are proliferating. Adequate coverage encounters difficult problems related to partiality, under specification, agents, and context dependency, which are signature features of information in nature, natural languages, and reasoning.

The session covers theoretical work, applications, approaches, and techniques for computational models of information, language (artificial, human, or natural in other ways), and reasoning. The goal is to promote computational systems and related models of thought, mental states, reasoning, and other cognitive processes.

Organizing Committee

Roussanka Loukanova, Institute of Mathematics and Informatics, Bulgarian Academy of Sciences, Sofia, Bulgaria
Sara Rodríguez, University of Salamanca, Salamanca, Spain

Program Committee

Benedikt Ahrens, School of Computer Science, University of Birmingham, Birmingham, UK
Krasimir Angelov, University of Gothenburg and Chalmers University of Technology, Gothenburg, Sweden
Wojciech Buszkowski, Faculty of Mathematics and Computer Science, Adam Mickiewicz University, Poznan, Poland
Marie Duzi, VSB-Technical University of Ostrava, Ostrava, Czech Republic
Antonín Dvořák, University of Ostrava, Ostrava, Czech Republic
Annie Foret, IRISA and University of Rennes 1, France
Håkon Robbestad Gylterud, University of Begen, Bergen, Norway
Ali Hürriyetoğlu, Koç University, Istanbul, Turkey
M. Dolores Jiménez López, GRLMC-Research Group on Mathematical Linguistics, Universitat Rovira i Virgili, Tarragona, Spain
Manfred Kerber, School of Computer Science, University of Birmingham, UK
Peter Koepke, Mathematisches Institut, Rheinische Friedrich-Wilhelms-Universität Bonn, Bonn, Germany
Stepan Kuznetsov, Steklov Mathematical Institute of RAS and HSE University, Moscow, Russia
Kristina Liefke, Ruhr-University Bochum, Bochum, Germany
Zhaohui Luo, Royal Holloway, University of London, London, UK
Richard Moot, Université de Montpellier and LIRMM-CNRS, Montpellier, France
Petra Murinová, Institute for Research and Applications of Fuzzy Modeling, University of Ostrava, Ostrava, Czech Republic
Rainer Osswald, Heinrich-Heine-Universität Düsseldorf, Düsseldorf, Germany
Christian Retoré, Université de Montpellier and LIRMM-CNRS, Montpellier, France
Frank Richter, Goethe University Frankfurt a.M., Frankfurt, Germany
Ana Paula Rocha, University of Porto, LIACC / FEUP, Porto, Portugal
Sylvain Salvati, Université de Lille, INRIA, CRIStAL UMR 9189, Lille, France
Milena Slavcheva, Bulgarian Academy of Sciences, Sofia, Bulgaria
Alexander Steen, Universität Greifswald, Institut für Mathematik und Informatik, Greifswald, Germany

Alexey Stukachev, Sobolev Institute of Mathematics, Novosibirsk State University, Novosibirsk, Russia
Satoshi Tojo, School of Information Science, Japan Advanced Institute of Science and Technology (JAIST), Japan
Jørgen Villadsen, Technical University of Denmark, Copenhagen, Denmark
Marek Zawadowski, Institute of Mathematics, University of Warsaw, Warsaw, Poland

Towards Ontology-Based End-to-End Domain-Oriented KBQA System

Anastasiia Zakharova(✉), Daria Sorokina, Dmitriy Alexandrov, and Nikolay Butakov

ITMO University, St. Petersburg, Russia
nastyazakharova.nz@gmail.com

Abstract. Knowledge Base Question-Answering (KBQA) systems have a number of benefits over Question-Answering over Text (QAT) systems: the ability to process complex, multi-hop questions and the absence of the need for answers containing texts. However, to develop a KBQA system, a knowledge base (KB) and a dataset for training is required. This fact is a significant drawback in the domain and non-English language conditions. We propose the end-to-end domain-oriented KBQA system concept consisting of two methods: filling in the ontology using weak-supervision methods based on the BERT model and KBQA training methods based on graph data. This approach requires only an ontology, raw texts, and a small amount of labeled data. We also analyze the robustness of the KBQA system in conditions of limited data.

Keywords: Question answering · BERT · Data mining

1 Introduction

Nowadays, there are two main approaches for building question-answering (QA) systems: QAT and KBQA. Recent work on QAT systems relies on transformers-based models for answer span extraction, utilizing context-question pairs as an input. Albeit the efficacy of such models, QAT systems are usually limited with the requirements of an answer-containing context, factoid questions nature, and lack of multi-hop reasoning. On the other hand, a KBQA-system copes with complex questions and does not require relevant texts, however, KB should be filled.

While a large amount of work is devoted to the creation of open-domain KBQA systems based on Freebase [3], DBpedia [1], and Wikidata [21] graphs, not enough attention is paid to domain-specific KBQA systems, which can be useful both for finding answers to knowledge within the company and for creating services and chatbots in various fields: cars, real estate, medicine, etc. However, to build a domain KBQA system, it is necessary to have both a KB and datasets for training in the form of a question-query-answer, a question-triplet/doublet-answer, or others. At the same time, system development in non-English domains poses a challenging task due to domain-specific data and trained models scarcity.

J. M. Machado et al. (Eds.): DCAI 2022, LNNS 585, pp. 37–47, 2023.
https://doi.org/10.1007/978-3-031-23210-7_4

We propose the concept of prototyping such systems in an end-to-end setting: from ontology and domain texts to a fully functioning KBQA system using a small amount of labeled data. It is important to note that ontology serves to structure and limit the knowledge that is needed to create a service or product. The proposed approach consists of two parts: ontology population and KBQA. The ontology population uses the RuBERT [13] model for entity recognition, relation classification, and QAT tasks to extract (subject; predicate; object) triplets from texts to add to the KB. Collection of the training data for aforementioned models is performed in a weakly supervised manner, which is described in Sect. 3.1.

The idea of the second part of the approach is to use KB triplets and pairs to generate a dataset for training the JointBERT [6] model, which allows learning from the intent classification and slot filling tasks. During the inference, a triplet/pair is extracted from a natural question, then the SPARQL generation module forms a query to get an answer from KB. We also conducted controlled experiments to test the robustness of the KBQA method depending on the number of mined triplets from the KB-filling method.

Our contributions in this work are the following:

- We propose an end-to-end concept for building a domain-specific KBQA system using nothing but an ontology, raw texts, and a few examples of labeled data.
- We propose weak-supervision methods for training data generation and conduct an experimental study on ontology population and KBQA tasks.
- We conduct experiments to analyze the robustness of the KBQA system in limited data conditions.

2 Related Work

2.1 Information Extraction

Information extraction (IE) methods can be divided into two groups based on the presence of ontology: OpenIE and Ontology population. OpenIE methods allow extracting an unlimited amount of entity and relation types. Whereas in ontology population tasks, there are entities, which are instances of class/subclasses, and relations between classes fixed.

Ontology population systems such as [9,15] leverage or generate a set of rules. However, these approaches require a lot of human involvement and do not scale for other domains. Supervised methods such as [2,25] lack data for training and often ask domain experts to accept or reject ontology candidates.

Supervised OpenIE methods include formulating tasks as a Sequence Labeling [19] or Text2Triple translation [16,22].

Some works are devoted to the development of weakly supervised IE systems. In paper [23], authors match Wikipedia articles with attribute-value pairs from infoboxes using heuristics. Researchers in works [18,26] train BiLSTM models on data derived from inferencing the OpenIE5 system on raw text.

The main advantage of our work lies in ontology population requiring minimal human supervision, described in Sect. 3.1, and a few hand-crafted samples, i.e. generated relation examples, one for each relation class. Moreover, the task becomes even more challenging as requires triplets extraction from non-English domain-specific texts and lack of existing IE systems such as OpenIE5 for deriving weak labels as in [18,26].

2.2 KBQA Methods

KBQA-systems commonly trained and validated on popular datasets: SimpleQuestions [5], ComplexWebQuestions [20], etc. The main part of these datasets is based on large-scale KBs: Freebase [3], DBpedia [1], and Wikidata [21]. In paper [11], natural questions represent a graph for retrieving answer's subgraph, whereas in [4], authors leverage similarity between question embedding and answer's subgraph embedding. In works [5,7] memory networks model questions-subgraph matching. Some approaches for extracting and linking entities and relations leverage CNN [8,24] or BiLSTM [10] architectures, or RL [14] methods. All the abovementioned methods use datasets for training, and some of them derive extracted entities from Freebase API and employ entity/relation-alias dictionaries [11]. Question generation approaches [4,5] construct pattern-based questions using KB, nevertheless, the model trains on a hand-crafted dataset.

Our work investigates the possibility of developing a KBQA-system in the low-resource data setting: we do not have a dataset and paraphrase database.

3 Methods

Proposed methods allow deploying KBQA-system having only ontology and few hand-crafted examples. A two-stage KBQA building procedure was developed. At the first step, the KB-filling method, in which raw text is utilized as a source for ontology population. At the second step, the KBQA method, which allows dataset construction from KB triples to train models for KBQA is developed and used to perform the KBQA task.

3.1 KB-Filling Methods

Knowledge base RDF-triples are stored in a $\{subject; predicate; object\}$ form. To collect such triples from raw texts, labeled data is required. Low-resource setting of non-English language automotive domain constrains the applicability of publicly available entity recognition and relation classification datasets, thus we propose a weakly supervised method for triplets extraction.

We demonstrate the diagram of the triplets extraction method in Fig. 1.

The extraction is performed as follows: (1) raw texts are passed into the entity recognition (ER) and relation classification (RelClf) models; (2) retrieved entity and relation (if exist) are used as the subject and predicate; (3) corresponding to the relation *golden template question*, which is described further, is filled with

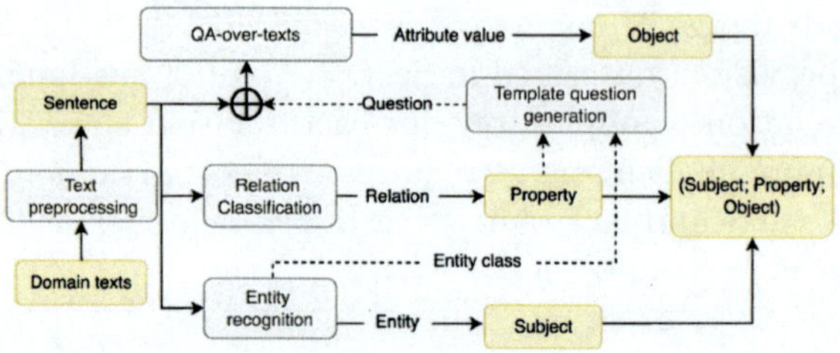

Fig. 1. KB-filling method

an entity class alias and passed into the QA system as the question. Input text is used as the context, and the model retrieves an object as an answer span.

Entity Recognition. To obtain data for an ER model training, the following method was proposed: (1) for each entity class an alias in natural language is created. We apply normalization technique at each source text and class alias, utilizing PyMorphy2 [12]; (2) we parse sentence tokens according to the Universal Dependencies scheme[1] and extract noun phrases used as *entity candidates*; (3) we embed *entity candidates* passing the sentences into the BERT model and cluster obtained contextualized vectors via the Agglomerative clustering algorithm implemented in scikit-learn [17]; (4) for each cluster, we count matches between entity aliases and text tokens, *entity candidates* in the most populated clusters are tagged according to prevalent entity class value with expert supervision. The expert skims over the extracted entity candidates in mostly populated clusters to approve correctly identified synonyms and discard the cluster otherwise.

Relation Classification. Training data for a RelClf model was obtained from the question-answer (QA) pairs from automotive domain forum described in Sect. 4.1. Weak-supervision data annotation method includes the following steps: (1) for each relation type the *golden template questions* (examples of relations with an empty entity class field) are created; (2) via the semantic textual similarity (STS) model[2] each text is compared with filled *golden templates*; (3) we filter texts with respect to obtained scores, and assign the relation label to the question according maximum STS-score value; (4) finally, we clean the expert answers and label them respectively to the paired question.

Question-answering-over-text.

Weakly-supervised Dataset. Section 4.1 provides the description of utilized data sources for QAT training data collection. The weak-supervision question-answer-context triples mining method consists of the following steps: (1) context ranking is performed according to the STS-score, introduced previously; (2) to create candidate spans, we split each context into n-grams, where $n \in \{1, 20\}$ with stride 1. Due to aforementioned domain and language specifics, we utilize

[1] https://github.com/deepmipt/DeepPavlov/blob/master/docs/features/models/syntaxparser.rst.

[2] https://huggingface.co/sentence-transformers/distiluse-base-multilingual-cased-v1.

BERT-score [27] and retrieve top-1 span as the answer; (3) additionally, the training data is enlarged by examples, where expert answers are imputed in the context at the random position and further used as answer span.

Reading Comprehension Dataset. Due to the factoid nature of the questions required for triples extraction, SberQuAD dataset, provided by Huggingface[3], is also used in an experimental study. Firstly, we fine-tune the QA model on clean SberQuAD samples, and then continue training the model with weakly-supervised data.

3.2 KBQA-Methods

As knowledge graphs question-answering is limited by stored set of entities and relations, the proposed KBQA-method illustrated in Fig. 2 implies not only natural language question processing, but also question generation based on knowledge graph's triplets and further datasets collection for both training and inference, respectively.

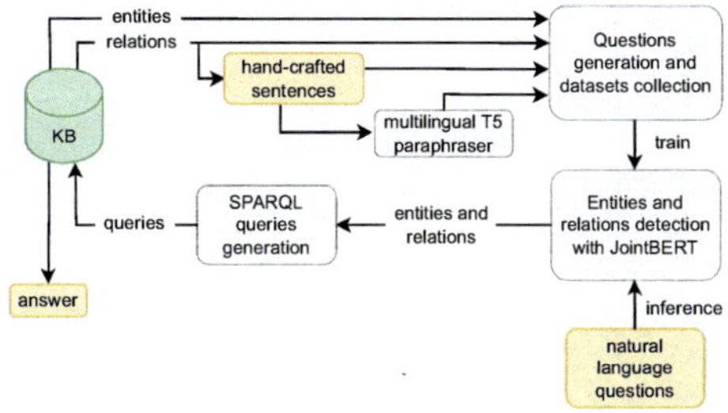

Fig. 2. KBQA method overview

Question Processing. Developed question-answering method considers KB entities and relations extraction as slot filling and intent detection. While there is a variety of approaches, joint models have proven efficacy, exploiting the relationship between two tasks. Therefore, JointBERT was fine-tuned on knowledge graph-based datasets and utilized to retrieve information, needed to form corresponding SPARQL queries and obtain answer.

Question Generation. While question processing requires entities and relations detection, question generation allows datasets collection for model training, utilizing knowledge graph's triplets. To perform a comparative research three sequential modules were applied and five datasets were generated. (1) Entities processing: *Triplets* dataset is represented by the set of samples in a $\{subject; predicate; object\}$ form, while other *(datasets: Duplets, Randomly placed duplets, Paraphrased context, Paraphrased relation)* incorporate either *subject* or *object*, simulating single-hop questions. (2) Relations processing: In addition, the relations of the knowledge graph were manually transformed

[3] https://huggingface.co/.

into domain language sentences *(datasets: Triplets, Duplets, Randomly placed duplets)*, and paraphrased with multilingual T5[4] versions *(datasets: Paraphrased context, Paraphrased relation)*. In case of *Paraphrased context* dataset, relations were enriched with natural language question syntax. Although paraphrased sentences enlarge the scope of the dataset, enhancing semantic parsing performance and robustness, imperfection of paraphrasing results in noisy data presence. (3) Entities and relations linking: To obtain relevant questions, retrieved from knowledge graph entities were canonically, in a $\{subject; predicate; object\}$ form *(datasets: Triplets, Duplets)* or randomly *(datasets: Randomly placed duplets, Paraphrased context, Paraphrased relation)* inserted in hand-crafted and paraphrased sentences. Examples of collected datasets are presented in Table 1.

Table 1. Collected datasets' examples

Dataset	Examples
Triplets	Renault Logan CO_2 emission rating 97 g/km
Duplets	Renault Logan CO_2 emission rating
Randomly placed duplets	CO_2 emission Renault Logan rating
Paraphrased context	What average gas emissions does Renault Logan have
Paraphrased relation	Renault Logan emissions of carbon dioxide

4 Results

Current section contains an experimental study investigating the system quality and robustness. We use Tesla V100 and Tesla T4 GPUs for our experiments.

4.1 Datasets

Ontology Population And KBQA Data Sources. An experimental study on RDF-triples extraction was performed on anonymized data parsed from an open-source, namely automotive question-answering forum and car and consumable items' reviews from two online aggregators. Training of the QA model comprised general-domain reading comprehension dataset SberQuAD. For the ontology population and method validation, we parse raw texts from articles on automotive domain websites. KBQA system relies on automotive-domain knowledge graph, which was populated with catalogues data from automotive website that comprise car characteristics and details' information, and knowledge graph collected via weak-supervision triples extraction.

Datasets. Weak-supervision methods shown in Sect. 3.1 for an entity recognition 25 K training samples were collected; for relation classification - 666 samples; for question-answering model - 95K triples. Validation datasets comprises 100, 86 and 88, respectively. Methods represented in Sect. 3.2 results in collection over 80 K training samples based on automotive-domain knowledge graph and validation was performed on 216 samples manually labeled.

[4] https://github.com/RussianNLP/russian_paraphrasers.

4.2 KB-Filling Methods

An experimental study on the ontology population is performed based on russian RuBERT-base-cased model, pretrained version of which given by Huggingface. For each subtask the best results are reported with respect the task evaluation metrics. Validation results are shown in Table 2.

Table 2. Triple extraction experimental results

Task	$Accuracy/Exactmatch$	$Precision$	$Recall$	$F1_{score}$
Relation classification	0.82	0.84	0.83	0.80
Entity recognition	0.92	0.65	0.63	0.64
Question-answering	0.63	–	–	0.75
Triplet extraction	–	0.63	0.62	0.63

Regarding the ontology population task, the precision and recall values are calculated according to the confusion matrix, where TP value is calculated as the number of triplets presented in a text and extracted by the system correctly; FP - number of extracted triples, which are not presented in a text; FN - number of missed by the system relations.

It can be seen that described methods allowed model to achieve reasonable performance on a weakly-supervised data. With regards to the question-answering task, an experimental study demonstrated that fine-tuning of the model trained on SberQuAD dataset with weakly supervised dataset significantly increases performance by 15% (from 0.65 to 0.75 in terms of F1-score). As for the triplets extraction, developed method retrieved more than 600 triples from domain-specific texts with noticeable precision and recall values.

Conducted human assessment of the method's predictions has provided several insights into the most common errors in a triplet extraction task. It was identified, that low-resource setting results in false positive triples extraction, where the entity type or the relation class were falsely retrieved by the trained models due to lack of comprehensive domain thesaurus, and scarcity of the negative examples for the relation classes, resulting in overfitting to the presence of "trigger" words in data. Moreover, in a question-answering model the bias towards a longer answer extraction has emerged due to the natural answer lengths distribution. Nevertheless, in current research, the prototype of the system was developed, thus mitigation of those limitations is in the scope of further work.

4.3 KBQA-Methods

The results of joint entities and relations detection are presented in Table 3 in terms of entity detection F1-score ($F1_{ent}$), relation detection accuracy (Acc_{rel}) and QA accuracy (Acc_{QA}).

Table 3. KBQA experimental results

Dataset		Test1			Test2	
	$F1_{ent}$	Acc_{rel}	Acc_{QA}	$F1_{ent}$	Acc_{rel}	Acc_{QA}
Triplets	0.28	0.40	0.01	0.27	0.45	0.03
Duplets	0.63	0.65	0.25	0.68	0.64	0.32
Randomly placed duplets	0.74	0.63	0.38	0.72	0.64	0.40
Paraphrased context	0.99	0.98	0.98	0.93	0.85	0.85
Paraphrased relation	0.94	0.81	0.81	0.93	0.80	0.79

The model that was trained on the *Triplets* dataset expectedly reveals the worst performance on the downstream tasks due to specific triplet structure and incorporation of both subject and object. Models that were trained on other datasets show quality enhancement, consequently benefit from single-hop question-answering simulation, non-structured positions of entities and relations in generated questions and context variability.

To obtain answers on the natural language questions posed, corresponded SPARQL queries were generated based on the retrieved entities and relations. In general, knowledge graph question-answering performance plummets in comparison to joint entities and relations detection results. The results of model trained on *Paraphrased context* and *Paraphrased relation* datasets increase question-answering accuracy up to relation detection (0.98 and 0.81 for $Test1$ and $Test2$, respectively) due to high quality entities detection.

Context paraphrase requires large-scale training data, capturing both semantic and syntactic variance, which could not be obtained without human supervision and expert knowledge. Considering low-resource setting constrains, to reduce the computational complexity KBQA-method based on relation paraphrase is applied.

5 System Robustness Experiments

To examine the robustness of proposed KBQA-method, we gradually decreased number of training dataset samples and analyzed corresponding performance reduction. Results obtained in an experimental study are demonstrated in Table 4. We report the same quality metrics, namely $F1_{ent}$ and Acc_{rel}, as in Sect. 4.3. Comparing models evaluation results, insignificant performance drop can be observed up to 5% of provided training samples utilization. While entity detection F1-score is revealed to be good enough even in few-shot setting, the quality of relation detection is highly sensitive to the lack of data variety. The results of $F1_{ent}$ can be explained by unified syntactic and semantic structure of entities within ontology classes.

Table 4. System robustness experimental results

Paraphrased relation dataset	$F1_{ent}$	Acc_{rel}	Paraphrased relation dataset	$F1_{ent}$	Acc_{rel}
100%	0.94	0.81	5%	0.91	0.79
70%	0.95	0.80	1%	0.84	0.63
50%	0.93	0.81	0.5%	0.78	0.52
30%	0.93	0.80	0.2%	0.71	0.25
10%	0.93	0.80	2 samples/rel	0.71	0.18

6 Conclusion

The paper proposes the concept of an end-to-end domain-oriented KBQA system. The automotive domain is being considered for its testing. Weak supervision ontology population method recovers more than 600 triplets with F1-score equal to 0.63. Moreover, additional weakly labeled data for the QAT module increases QAT quality by 15%. The KBQA method with context paraphrase ($F1 = 0.98/0.85$) and relationship paraphrase ($F1 = 0.81/0.79$) shows a strong increase in quality compared to duplets and randomly placed duplets. Experiments testing the robustness of the KBQA method show that 5% of the data still has high-quality indicators (−3.2%/ -2.5%) for entity recognition and relationship classification), and even 0.5% of the data has acceptable quality indicators (−17%/-36%). Thus, the proposed concept has the potential to exist within the framework of prototyping KBQA systems in conditions of data limitations. Further research should be devoted to the complete combination of the two methods and the reuse of models or collaborative learning.

Acknowledgements. This research is financially supported by The Russian Science Foundation, Agreement #20-11-20270.

References

1. Auer, S., Bizer, C., Kobilarov, G., Lehmann, J., Cyganiak, R.: Dbpedia: a nucleus for a web of open data. In: The semantic web, pp. 722–735. Springer (2007)
2. Ayadi, A., Samet, A., de Beuvron, F.d.B., Zanni-Merk, C.: Ontology population with deep learning-based nlp: a case study on the biomolecular network ontology. Proced. Comput. Sci. 572–581 (2019)

3. Bollacker, K., Evans, C., Paritosh, P., Sturge, T., Taylor, J.: Freebase: a collaboratively Created graph database for structuring human knowledge (2008)
4. Bordes, A., Chopra, S., Weston, J.: Question answering with subgraph embeddings. arXiv preprint arXiv:1406.3676 (2014)
5. Bordes, A., Usunier, N., Chopra, S., Weston, J.: Large-scale simple question answering with memory networks. CoRR abs/1506.02075 (2015)
6. Chen, Q., Zhuo, Z., Wang, W.: Bert for joint intent classification and slot filling. ArXiv preprint arXiv:1902.10909 (2019)
7. Chen, Y., Wu, L., Zaki, M.J.: Bidirectional attentive memory networks for question answering over knowledge bases. In: NAACL HLT 2019 - Proceedings of the Conference pp. 2913–2923 (2019). https://doi.org/10.18653/v1/n19-1299
8. Dong, L., Wei, F., Zhou, M., Xu, K.: Question answering over freebase with multicolumn convolutional neural networks. In: Proceedings of the 53rd Annual Meeting of the ACL. pp. 260–269 (2015). https://doi.org/10.3115/v1/p15-1026
9. Faria, C., Serra, I., Girardi, R.: A domain-independent process for automatic ontology population from text. Sci. Comput. Programm. 26–43 (2014). https://doi.org/10.1016/j.scico.2013.12.005
10. Hao, Y., Zhang, Y., Liu, K., He, S., Liu, Z., Wu, H., Zhao, J.: An end-to-end model for question answering over knowledge base with cross-attention combining global knowledge. In: ACL 2017, Proceedings of the Conference), pp. 221–231 (2017)
11. Hu, S., Zou, L., Yu, J.X., Wang, H., Zhao, D.: Answering natural language questions by subgraph matching over knowledge graphs. IEEE Trans. Knowl. Data Eng. 824–837 (2017)
12. Korobov, M.: Morphological analyzer and generator for russian and ukrainian languages. In: Analysis of Images. Social Networks and Texts, pp. 320–332. Springer, Communications in Computer and Information Science (2015)
13. Kuratov, Y., Arkhipov, M.: Adaptation of deep bidirectional multilingual transformers for russian language. arXiv preprint arXiv:1905.07213 (2019)
14. Lan, Y., Wang, S., Jiang, J.: Knowledge base question answering with topic units. Int. Joint Conf. Artif. Intell. 5046–5052 (2019)
15. Makki, J., Alquier, A.M., Prince, V.: Ontology population via NLP techniques in risk management. Int. J. Human. Soc. Sci. 212–217 (2008)
16. Paolini, G., Athiwaratkun, B., Krone, J., Ma, J., Achille, A., Anubhai, R., Santos, C.N.D., Xiang, B., Soatto, S.: Structured prediction as translation between augmented natural languages. arXiv preprint arXiv:2101.05779 (2021)
17. Pedregosa, F., Varoquaux, G., Gramfort, A., Michel, V., Thirion, B., Grisel, O., Blondel, M., Prettenhofer, P., Weiss, R., Dubourg, V.: Scikit-learn: machine learning in python. J. Mach. Learn. Res. **12**, 2825–2830 (2011)
18. Roy, A., Park, Y., Lee, T., Pan, S.: Supervising unsupervised open information extraction models. In: EMNLP-IJCNLP 2019, Proceedings, pp. 728–737 (2019)
19. Stanovsky, G., Michael, J., Zettlemoyer, L., Dagan, I.: Supervised open information extraction. In: NAACL HLT 2018 - Proceedings of the Conference, pp. 885–895 (2018). https://doi.org/10.18653/v1/n18-1081
20. Talmor, A., Berant, J.: The web as a knowledge-base for answering complex questions. NAACL HLT 2018–2018 Conference - Proceedings of the Conference, pp. 641–651 (2018). https://doi.org/10.18653/v1/n18-1059
21. Vrandečić, D., Krötzsch, M.: Wikidata: a free collaborative knowledgebase. Commun. ACM 78–85 (2014). https://doi.org/10.1145/2629489
22. Wang, C., Liu, X., Chen, Z., Hong, H., Tang, J., Song, D.: Zero-Shot Information Extraction as a Unified Text-to-Triple Translation. ArXiv, pp. 1225–1238 (2021). https://doi.org/10.18653/v1/2021.emnlp-main.94

23. Wu, F., Weld, D.: Open information extraction using wikipedia. In: Proceedings of ACL, pp. 118–127 (2010)
24. Yin, W., Yu, M., Xiang, B., Zhou, B., Schütze, H.: Simple question answering by attentive convolutional neural network. In: Conference on Computational Linguistics, Proceedings of COLING 2016: Technical Papers, pp. 1746–1756 (2016)
25. Zeng, D., Liu, K., Lai, S., Zhou, G., Zhao, J.: Relation classification via convolutional deep neural network. In: Proceedings of COLING 2014, the 25th International Conference on Computational Linguistics: Technical Papers, pp. 2335–2344 (2014)
26. Zhan, J., Zhao, H.: Span model for open information extraction on accurate corpus. In: Proceedings of the AAAI Conference on Artificial Intelligence, pp. 9523–9530 (2020). https://doi.org/10.1609/aaai.v34i05.6497
27. Zhang, T., Kishore, V., Wu, F., Weinberger, K.Q., Artzi, Y.: Bertscore: evaluating text generation with bert. In: International Conference on Learning Representations (2020)

TFEEC: Turkish Financial Event Extraction Corpus

Kadir Şinas Kaynak(✉) and Ahmet Cüneyd Tantuğ

Istanbul Technical University, 34469 Maslak, Istanbul, Turkey
{kaynakk19,tantug}@itu.edu.tr

Abstract. Event extraction from the news is essential for making financial decisions accurately. Therefore, it has been researched in many languages for a long time. However, to the best of our knowledge, no study has been conducted in the domain of Turkish financial and economic text mining. To fill this gap, we have created an ontology and presented a well-defined and high-quality company-specific event corpus of Turkish economic and financial news. Using our dataset, we conducted a preliminary evaluation of the event extraction model to serve as a baseline for further work. Most approaches in the event extraction domain rely on machine learning and require large amounts of labeled data. However, building a training corpus with manually annotated events is a very time-consuming and intensive process. To solve this problem, we tried active learning and weak supervision methods to reduce human effort and automatically produce more labeled data without degrading machine learning performance. Experiments on our dataset show that both methods are useful. Furthermore, when we combined the manually annotated dataset with the automatically labeled dataset and used it in model training, we demonstrated that the performance increased by %2,91 for event classification, %13,76 for argument classification.

Keywords: Event extraction · Semi-supervised · Weak supervision · Active learning · Corpus generation

1 Introduction

News in the finance and economy domains attracts everyone's attention, whether they are investors or not. News provides information on many subjects, such as market analysis, financial instruments, corporate events, regulatory policies, and decisions that may affect the markets. Hence, people benefit from the news in order to make better decisions and get solid future predictions. However, thousands of news articles are reported by hundreds of sources every day, and it is not easy to follow them. Automatic event extraction from the news allows us to do a lot of work in a short time and makes our lives easier. The task that tackles this problem in the Natural Language Processing (NLP) field is named event extraction [4]. Briefly, Event Extraction (EE) allows us to automatically

J. M. Machado et al. (Eds.): DCAI 2022, LNNS 585, pp. 49–58, 2023.
https://doi.org/10.1007/978-3-031-23210-7_5

determine the expressions that represent the event, the arguments involved in the event, and their relationships. The first studies carried out on this task were generally applied with the pattern matching methods [23]. In such a rule-based system, domain experts design and build rules and templates, and then the events and their arguments are extracted by matching these rules in the running text. The accuracy of the method is high due to the well-defined templates created by the experts. However, since large-scale event design is costly and time-consuming, machine learning-based techniques have been widely used in the following studies. Although traditional machine learning algorithms such as support vector machines [1], maximum entropy [2] and deep learning approaches such as convolutional neural networks [20], recurrent neural networks [18], and graph neural networks [17] are used separately in different studies, the basic idea of machine learning approaches is nearly the same. The idea is to learn classifiers from training data and apply them for event extraction from news. The classifiers are trained on the ground truth labeled set and are used to extract events from the text. However, training a high-performing classifier requires a large dataset, and generating a labeled dataset is an expensive, time-consuming, and labor-intensive process. Recently, many studies on other NLP related tasks report state-of-the-art results by using semi-supervised training techniques [5]. In these studies, minimal human annotation efforts are employed to build a large-scale dataset to minimize the disadvantages of manual annotation labor. On the other hand, most of the previous event extraction studies were concentrated on languages like English [10], Chinese, Arabic [4] and Dutch [12], for which economic and financial event labeled datasets already exist. In the specific case of Turkish, we are not aware of either a specific dataset nor a system for economic and financial event extraction tasks.

In this study, we report three main contributions. Our first contribution is an ontology consisting of 25 types and 104 subtypes by analyzing the ontologies in other languages and the company-specific Turkish news collected from various sources. Using this ontology, we manually annotated the event triggers and arguments in 600 news articles and presented them as a gold standard corpus of Turkish financial and economic news articles to provide a labeled dataset that can be used in training supervised machine learning models. The second contribution is to provide a baseline result of Turkish financial event extraction. By using this dataset, we trained and build a model as a preliminary evaluation of our dataset. The performance of this model will serve as a baseline for further studies. Lastly, we conduct experiments by employing Active Learning (AL) and Weak Supervision (WS) methods to increase the amount of labeled data and minimize manual annotation efforts. We expanded our dataset by combining the new labeled data obtained automatically with our gold set, and experimental results show that the model trained on this expanded dataset achieved better performance.

2 Related Work

2.1 Event Extraction Corpora

Annotation is the process of enabling the machine learning model to make the correct prediction by showing the result you want it to predict. The training data generated by this process guides the models to accurately understand the given tasks and make the desired predictions. Therefore, generating a corpus is essential for model success. But this may be more difficult than it seems. In this section, we will review the relevant studies that have been done.

Normally, the dataset to be generated is annotated manually by experts with domain knowledge. But this is time-consuming and intensive work for them, so early studies on corpus generation have always been small and less comprehensive. For example, the ACE 2005 corpus [4] consists of a total of 599 documents, including English, Arabic and Chinese languages, and includes 8 events and 33 subtypes. The SENTIVENT corpus [10] includes approximately 6,203 events in 288 documents in English, and 18 event types and 64 subtypes were defined in this study. We have adapted to event trigger and argument annotating processes in other studies in this domain. However, since the types and subtypes used in those studies were not sufficient for us, we expanded them with new event types and subtypes that we determined according to the news we read.

2.2 Data Expansion

Machine learning algorithms need large amounts of data to learn events, complex patterns, and to generalize better. This is mostly because more data helps the model to better understand the underlying patterns and increase the chances of making successful predictions. Normally, the traditional approach is to hire experts to label the data, but if the dataset to be annotated is large, then the process becomes very expensive and difficult. To avoid data sparseness, we can develop models that use small labeled datasets or dictionaries and get more labeled data. In this section, we will review methods to obtain more labeled data.

Active Learning: The data given to build a good model does not have the same effect. The aim of the AL approach is to pick the most valuable samples in terms of information rather than all available data. With these selective annotations, it can achieve the same or better results as more annotations. AL has been successfully applied to many tasks, such as named entity recognition [7], text classification [8], part of speech tagging [24], parsing [22], word sense disambiguation [28], and even event extraction [13]. We also evaluated AL in this study as it helps us prioritize the data we will process. Unlike other studies, we made experiments by including weak supervision functions in the active learning cycle.

Transfer Learning: Normally, learning from scratch is performed for machine learning. However, since it will be possible and advantageous to use what is learned in one task or topic to another, the information obtained from the source tasks is used for the solution of the target task. With learning transfer, models that show higher success and learn faster with less training data are obtained by using previous knowledge. For this reason, it has been preferred in many event extraction studies [9,19]. However, we could not use this method since there is no previous study on Turkish financial events that we can take as a reference.

Weak Supervision: This approach allows existing unorganized or imprecise information sources to provide indications to label unlabeled data. In order to obtain labeled data using WS, researchers can develop labeling functions [15] using resources such as knowledge bases [16], heuristics [11], and with help from domain experts. This makes it possible to leverage large amounts of cheaper, unlabeled data, and many studies use weak labels for model training, although they are often noisy and poor quality. Various systems have been developed [14,21] that allow us to obtain label sets of unlabeled data by writing small code snippets. In the Snorkel [21] tool, which is one of these systems, we received labeled data using the labeling functions we developed specifically for this task.

3 Corpus Construction

We have created a corpus that will enable us to extract events from Turkish finance and economy news articles. For this study, we crawled company news published on www.borsagundem.com between 2010 and 2022. The corpus described in this paper consists of 34,746 articles, containing 323,945 sentences and 6,180,243 tokens in total.

Before proceeding to the annotation step, we preprocessed the documents collected from news sites. In this process, we run a model [6,26] that labels named entity recognition tags and temporal expressions in sentences. These labels, which we obtained as a result of the model, will be used in the next step of automatic labeling. After completing the preprocessing step, we randomly selected 600 of these documents and labeled their event triggers and event arguments. Our annotation was done at the sentence level, and the WebAnno annotation tool [27] was used throughout the process.

Ontology allows us to define concepts within the field, the relationships between them and their properties. The purpose of creating an ontology for the detection of economic events is to allow the sharing of knowledge in this field, to ensure common understanding, and to support its reuse and development. At the beginning of the study, we reviewed other papers studying event extraction in the financial and economic domains [4,10]. Then we read numerous news articles from companies. We made updates to the set obtained from other studies according to the frequency and cohesion of the events in the news. As a result, we ended up with an ontology with 25 event types and 104 subtypes.

The list of event types, subtypes, and arguments involved in the event is available from https://github.com/kadirsinas/TFEEC. In addition to this, there are frequency distribution of event types extracted from the documents, statistical information about the dataset, instructions on how to markup, and sample screenshots.

The labeling processes in this study were made by the author (non-expert) of the article. A label on the gold standard should represent common understanding, and having only one annotator can be biased. Therefore, this issue will be discussed in the next study.

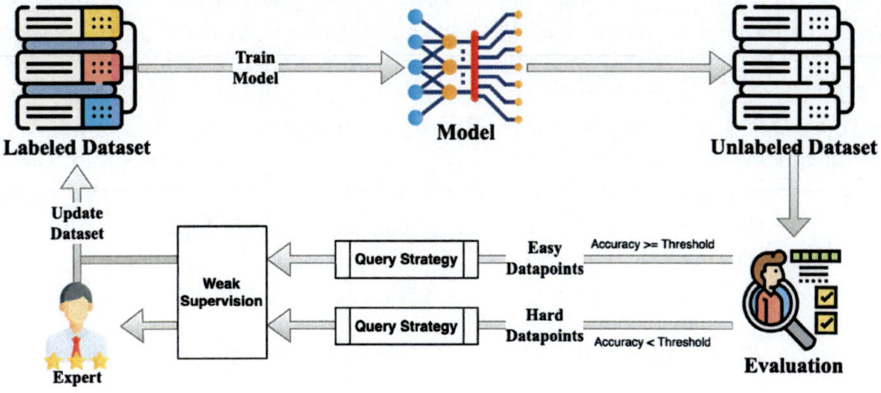

Fig. 1. System architecture.

4 Data Expansion Methodology

Our goal is to automatically obtain more high-quality labeled data by using the existing labeled dataset. In this way, the dataset that we obtain can be used to improve the performance of the event extraction models. We used AL and WS techniques to achieve this. In this section, we will discuss how we use them.

AL is used to intelligently pick data when the amount of data is huge, to achieve a high level of accuracy with fewer training labels. We also used it to prioritize samples from the unlabeled dataset to have the greatest impact in training a supervised model. We tried strategies such as least confidence, max confidence, margin sampling, and entropy selection to select the most informative queries.

WS is used to label unlabeled data using limited or imprecise sources. This approach simplifies costly and laborious work. Therefore, in this study we used the Snorkel tool [21] to implement WS and created two functions for it: *eventCheckFunc* and *labellingFunc*. The *labellingFunc* function gives the tags predicted by the trained classifier and name entity tags to Snorkel's generative model and integrates these inputs, so it can predict the correct class. The *eventCheckFunc* function uses a list of keywords we have defined to predict whether an event has occurred in a sentence. If it detects an event, it returns the result of the *labellingFunc* function. If it does not find an event, then assigns empty tags.

To benefit from the strengths of WS and AL, we also tried adding the WS functions to the AL loop. Figure 1 shows the general architecture of this method. This process starts with 2 datasets: labeled and unlabeled. The model is trained using the labeled dataset, and then predictions are made on the data in the unlabeled dataset. As a result of this, predictions that are above a certain value follow the Easy Data Points path, while others follow the Hard Data Points path. Different query strategies have been determined for these paths. Strategies that select reliable data, such as the maximum confidence strategy, are used for estimates above the threshold, as these data will be added to the labeled dataset with the labels assigned by the model, so it is important that the quality is not compromised. Those that went the other way were samples that couldn't exceed the threshold; queries like least confidence were used to find the most unstable data, and these samples were human-labeled to have the biggest impact on the model's performance. The results from the query strategies come to the WS module. The WS functions developed for use in this module can directly assign a label to the data or leave the annotation to the human. In the final stage, the annotated data is included in the labeled dataset, and this cycle continues until there is no unlabeled data left.

We consider the event extraction problem in two stages. The first stage is event classification (EC), at this stage, the event triggers in the text are determined, and the event type is predicted. The second stage, argument classification (AC), aims to extract the arguments from the text if the event is extracted in the first stage. The life cycle is the same for both phases. The only difference in AC is that the event type labels obtained from EC were also added to the WS process.

5 Evaluations

In the first part, we investigated the feasibility of the methods we tried and analyzed the outputs, finally showing whether the methods were effective in obtaining more labeled data and resulting in human effort reduction. In the second part, we compared the aforementioned datasets.

5.1 Event Extraction Cycle

Before all the work is done in this section, we decided to create a baseline. By this way, we will be able to make our comparisons and present them as a baseline model result for future studies. Therefore, we divided the hand-labeled dataset as 80% training 20% test set. Then we use a powerful general-purpose pretrained model [25] and fine-tune this network on our training dataset instead of training the models from scratch. This model allows us to make token-level predictions. We use this because the event extraction task is treated as a token classification problem. The parameters of the models used in the study are as follows: The learning rate is 2e-5, the sequence length is 128, the batch size per device during training is 3, the batch size for evaluation is 3, the total number of training epochs is 1, and the strength of weight decay is 0.01.

Table 1. Baseline model results.

Model	F1 Score	
	Event classification	Argument classification
BERT	85,89	65,81
ConvBERT	80,63	61,86
DistilBERT	78,57	58,16
ELECTRA	80,33	60,28
CRF	65,82	49,22

The models used and their macro-averaged F1-scores can be seen in the Table 1. When we examine the results, we can say that the BERT [3] model gives the best results for both classification tasks. These results were recorded as 85,89% for EC and 65,81% for AC. After observing the baseline performance results, we used the same test set used for the baseline model to measure the performance of other methods. The results of our other studies are shown in the Table 2 for both EC and AC.

Table 2. Results for classification tasks.

Tas	Methods	Training		Test	F1 Score	Change in percentage	Human effort saving
		Total number of sentence	Manually annotated sentence	Total number of sentence			
EC	**Baseline**	5 K	5 K	1 K	85,89		
	AL	3,75 K	3,75,K	1 K	85,57	−0,32%	25%
	AL + WS	5 K	2,3 K	1 K	86,51	+0,7%	54%
	WS	322 K	5 K	1 K	88,39	+2,91%	98%
AC	**Baseline**	5 K	5 K	1 K	65,81		
	AL	3,5 K	3,5 K	1 K	65,77	−0,06%	30%
	AL + WS	5 K	2 K	1 K	67,34	+2,32%	60%
	WS	322 K	5 K	1 K	74,87	+13,76%	98%

The first column *[Methods]* in the tables indicates which method we use. The second column *[Total Number of Sentence]* shows the total amount of data used in model training, while the third column *[Manually Annotated Sentence]* shows the amount of human-labeled data within this total amount of data. The fourth column *[Total Number of Sentence]* shows the number of data in the dataset where the trained model will be tested. The fifth column *[F1 Score]* contains the macro-averaged F1-scores of the methods. In the sixth column *[Change in Percentage]*, it is given what percentage of change that method provides compared to the baseline method. In the last column *[Saved Human Effort]*, it is shown what percentage of human effort is gained when working with the aforementioned method. To interpret the results, we trained a model using all human-labeled data as the baseline and took its results as the basis.

AL does not produce us labeled data as it does in WS, the main purpose of using active learning is to train the model using the minimum amount of labeled data that can reach the basic model performance. We can see that the performance of the baseline model we trained with 5 K data can be achieved with very small differences using less data (3.75 K for EC and 3.5 K for AC).

When we apply the method we proposed using AL and WS together, only 2 K of manually annotated data is used, and the remaining 3 K of data is automatically labeled. Using AL and WS together not only saves effort but also improves performance. When we compare this method with the baseline model, we see that the performance has increased even though amount of data in the training sets is the same. This was made possible by AL prioritizing valuable data and WS producing quality estimates.

Building on the good results of WS, we also wanted to label the unlabeled data pool we scraped from the news sites and evaluate the results. For model training, we used 5 K manually annotated data and 317 K annotated by WS. As a result, we see an increase of 2.91% for EC, and an increase of 13,76% for AC.

Table 3. Corpus statistics.

	ACE05	SENTIVENT	TFEEC manually annotated	TFEEC automatically annotated
Documents	599	288	600	34.146
Sentences	18.927	6.883	6.249	317.696
Tokens	303.000	170.398	117.206	6.063.037
Events	5.055	6.203	7.824	303.480
Arguments	6.040	13.675	9.148	377.011
Type	8	18	25	25
SubType	33	64	104	104

5.2 Dataset

In this section, we will provide information about the datasets that we created manually and obtained automatically with two of the most important studies in this field. In our study, we manually labeled 600 documents in total, and separated 500 of them as training sets and 100 as test sets. When we look at the dataset obtained automatically, we can see that the number of documents is about 34 K, which is 50 times more than the dataset we marked manually. Corpus names and statistical information can be seen in Table 3. These numbers make our study comparable in size to important studies in this area.

6 Conclusion

Event extraction is an important task in the natural language processing world and is used to capture real-world changes, classify them by event type, and identify their event arguments. Event extraction corpus in the financial and economic domains are available in various languages, but as far as we know no corpus exists in Turkish.

In this paper, we presented a well-defined, high-quality Turkish dataset that provides a rich resource on events to be used as a training data for supervised models in economic and financial applications, and we also provided a preliminary study of Turkish event extraction. To expand this dataset and reduce human annotation effort without degrading quality, we explored AL and WS methods, as a result of this we have increased the hundreds of annotated documents in the dataset to tens of thousands of annotated documents.

Our experimental results demonstrate that the method applied not only produces high-quality labeled data, but also increases model performance. The public dataset can be accessed at https://github.com/kadirsinas/TFEEC.

References

1. Chen, C., Ng, V.: Joint modeling for Chinese event extraction with rich linguistic features. In: Proceedings of COLING 2012, pp. 529–544 (2012)
2. Chieu, H.L., Ng, H.T.: A maximum entropy approach to information extraction from semi-structured and free text. Aaai/iaai **2002**, 786–791 (2002)
3. Devlin, J., Chang, M.W., Lee, K., Toutanova, K.: Bert: pre-training of deep bidirectional transformers for language understanding. ArXiv preprint arXiv:1810.04805 (2018)
4. Eaton, J., Gaubitch, N.D., Moore, A.H., Naylor, P.A.: The ace challenge-corpus description and performance evaluation. In: 2015 IEEE Workshop on Applications of Signal Processing to Audio and Acoustics (WASPAA), pp. 1–5. IEEE (2015)
5. Ferguson, J., Lockard, C., Weld, D.S., Hajishirzi, H.: Semi-supervised event extraction with paraphrase clusters. ArXiv preprint arXiv:1808.08622 (2018)
6. Güneş, A., Tantuğ, A.C.: Turkish named entity recognition with deep learning. In: 2018 26th Signal Processing and Communications Applications Conference (SIU), pp. 1–4 (2018). https://doi.org/10.1109/SIU.2018.8404500
7. Hachey, B., Alex, B., Becker, M.: Investigating the effects of selective sampling on the annotation task. In: Proceedings of the Ninth Conference on Computational Natural Language Learning (CoNLL-2005), pp. 144–151 (2005)
8. Hoi, S.C., Jin, R., Lyu, M.R.: Large-scale text categorization by batch mode active learning. In: Proceedings of the 15th international conference on World Wide Web, pp. 633–642 (2006)
9. Huang, L., Ji, H., Cho, K., Voss, C.R.: Zero-shot transfer learning for event extraction. ArXiv preprint arXiv:1707.01066 (2017)
10. Jacobs, G., Hoste, V.: Sentivent: enabling supervised information extraction of company-specific events in economic and financial news. Language Resources and Evaluation, pp. 1–33 (2021)
11. Karamanolakis, G., Mukherjee, S., Zheng, G., Awadallah, A.H.: Self-training with weak supervision. ArXiv preprint arXiv:2104.05514 (2021)
12. Lefever, E., Hoste, V.: A classification-based approach to economic event detection in dutch news text. In: 10th International Conference on Language Resources and Evaluation (LREC), pp. 330–335. ELRA (2016)
13. Liao, S., Grishman, R.: Using prediction from sentential scope to build a pseudo co-testing learner for event extraction. In: Proceedings of 5th International Joint Conference on Natural Language Processing. pp. 714–722 (2011)

14. Lison, P., Barnes, J., Hubin, A.: skweak: Weak supervision made easy for NLP. ArXiv preprint arXiv:2104.09683 (2021)
15. Lison, P., Hubin, A., Barnes, J., Touileb, S.: Named entity recognition without labelled data: A weak supervision approach. ArXiv preprint arXiv:2004.14723 (2020)
16. Mintz, M., Bills, S., Snow, R., Jurafsky, D.: Distant supervision for relation extraction without labeled data. In: Proceedings of the Joint Conference of the 47th Annual Meeting of the ACL and the 4th International Joint Conference on Natural Language Processing of the AFNLP, pp. 1003–1011 (2009)
17. Nguyen, T., Grishman, R.: Graph convolutional networks with argument-aware pooling for event detection. In: Proceedings of the AAAI Conference on Artificial Intelligence, vol. 32 (2018)
18. Nguyen, T.H., Cho, K., Grishman, R.: Joint event extraction via recurrent neural networks. In: Proceedings of the 2016 Conference of the North American Chapter of the Association for Computational Linguistics: Human Language Technologies, pp. 300–309 (2016)
19. Nguyen, T.H., Fu, L., Cho, K., Grishman, R.: A two-stage approach for extending event detection to new types via neural networks. In: Proceedings of the 1st Workshop on Representation Learning for NLP, pp. 158–165 (2016)
20. Nguyen, T.H., Grishman, R.: Event detection and domain adaptation with convolutional neural networks. In: Proceedings of the 53rd Annual Meeting of the Association for Computational Linguistics and the 7th International Joint Conference on Natural Language Processing (Volume 2: Short Papers), pp. 365–371 (2015)
21. Ratner, A., Bach, S.H., Ehrenberg, H., Fries, J., Wu, S., Ré, C.: Snorkel: Rapid training data creation with weak supervision. In: Proceedings of the VLDB Endowment. International Conference on Very Large Data Bases, vol. 11, p. 269. NIH Public Access (2017)
22. Reichart, R., Rappoport, A.: Self-training for enhancement and domain adaptation of statistical parsers trained on small datasets. In: Proceedings of the 45th Annual Meeting of the Association of Computational Linguistics, pp. 616–623 (2007)
23. Riloff, E., et al.: Automatically constructing a dictionary for information extraction tasks. In: AAAI, vol. 1, pp. 2–1. Citeseer (1993)
24. Ringger, E., McClanahan, P., Haertel, R., Busby, G., Carmen, M., Carroll, J., Seppi, K., Lonsdale, D.: Active learning for part-of-speech tagging: Accelerating corpus annotation. In: Proceedings of the Linguistic Annotation Workshop, pp. 101–108 (2007)
25. Schweter, S.: Berturk - Bert models for Turkish. https://doi.org/10.5281/zenodo.3770924 (2020)
26. Uzun, A., Tantuğ, A.C.: Itutime: Turkish temporal expression extraction and normalization. In: International Symposium on Distributed Computing and Artificial Intelligence, pp. 74–85. Springer (2021)
27. Yimam, S.M., Gurevych, I., de Castilho, R.E., Biemann, C.: Webanno: A flexible, web-based and visually supported system for distributed annotations. In: Proceedings of the 51st Annual Meeting of the Association for Computational Linguistics: System Demonstrations, pp. 1–6 (2013)
28. Zhu, J., Hovy, E.: Active learning for word sense disambiguation with methods for addressing the class imbalance problem. In: Proceedings of the 2007 Joint Conference on Empirical Methods in Natural Language Processing and Computational Natural Language Learning (EMNLP-CoNLL), pp. 783–790 (2007)

Special Session on Intelligent Systems Applications (ISA)

Currently, due to the climate change, the aging of the population and the upcoming requirements of the industry and engineering, new challenges become of more interests and need to be achieved. Thus, minimizing the energy consumption, reducing the air pollution and emissions of greenhouse gases, improving the quality of life for human beings, or even achieving the industry requests are becoming of high interest and need to be taken into account by the industrial and academic personal, as they are in charge of the research and improvements that will define the future.

In consequence, the field of building management, assistive technologies, the applied engineering, and, in general terms, the industry needs to engage with the different upcoming techniques, which are developed for the purpose of achieving the previously mentioned goals. Until now, the traditional techniques have performed very efficiently with the current demands, nevertheless new advanced techniques are necessary for future and upcoming requests.

Organizing Committee

Denial of Service Attack Detection Based on Feature Extraction and Supervised Techniques

Álvaro Michelana[1(✉)], José Aveleira-Mata[2], Esteban Jove[1], Héctor Alaiz-Moretón[2], Héctor Quintián[1], and José Luis Calvo-Rolle[1]

[1] Department of Industrial Engineering, University of A Coruña, CTC, CITIC, Ferrol, A Coruña, Spain
{alvaro.michelena,esteban.jove,jlcalvo}@udc.es

[2] Department of Electrical and Systems Engineering, University of León, León, Spain
{jose.aveleira,hector.moreton}@unileon.com

Abstract. Internet of Things systems (IoT) is expanding exponentially, providing expanded services in different environments. The wide variety of these systems makes security an increasingly important challenge, several Malware, such as Mirai or Dark Nexus, demonstrate an increase in attacks based on IoT. One of the most used protocols in the application layer is the Message Queuing Telemetry Transport (MQTT), these systems can be attacked by Denial of Service attacks. This paper presents a framework for detecting MQTT protocol attacks based on automatic learning, using a dataset formed by all the network traffic generated in an environment that uses an IoT system with the MQTT protocol on which several DoS attacks are performed.

Keywords: MQTT · Cybersecurity · Classifiers · DoS

1 Introduction

IoT systems are being extended in various areas such as hospitals, homes, vehicles and industries. IoT adds functionalities to objects, with sensors and actuators. This allows observing, detecting and calculating for a more accurate decision making. IoT system devices require special capabilities, such as small size, low power consumption and scalable increase in number without saturating networks. For this they use specific lightweight protocols or other types of networks.

The most commonly used protocols in IoT systems are:

- In the low-level 6Lowpan or ZigBee using IEE802.15.4 and LoRA in sub-gigahertz radio frequency.
- In the application layer such as MQTT, CoAP, XMPP and DDS.

Since IoT is a complex, distributed and heterogeneous system by nature, new challenges in cybersecurity arise.

J. M. Machado et al. (Eds.): DCAI 2022, LNNS 585, pp. 61–70, 2023.
https://doi.org/10.1007/978-3-031-23210-7_6

Mirai that compromised 400,000 IoT devices and performed a massive DDoS attack that inflicted a loss of hundreds of millions of dollars to enterprises. Although, Mirai emerged in 2016, it is currently still active performing DDoS attacks.

Recently, in April 2020, a new botnet malware attack against IoT devices called Dark Nexus detected by cybersecurity researchers. It exploits vulnerable IoT devices to perform a DDoS attack. So far, the attack comprises about 1372 bots [13].

In order to address security in IoT systems, machine learning techniques such as classification, prediction, regression and the like are used, as well as deep learning techniques based on artificial neural networks [2]. These techniques are trained with different datasets containing network attacks. The most commonly used datasets are the NSL-KDD [16], or the AWID [18].

The current work is focused on the detection of DoS attacks to MQTT protocol, using supervised machine learning techniques, using a dataset generated in an environment that uses the MQTT with DoS attacks, to focus on the vulnerabilities of this protocol.

The present document is structured as follows: in Sect. 2 the case of study is presented. Then, Sect. 3 provides a detailed description of the machine learning methods used. The experiments and its results are shown in Sect. 4. Finally, the conclusions and future work are exposed in Sect. 5.

2 Case Study

Denial-of-service (DoS) is an attack for making an online service unavailable by flooding, based on high traffic levels from many hundreds or thousands of PCs in a short period of time. Attackers generate large volumes of traffic overload the server until the server cannot respond to all requests, as result, users cannot access the services of the server. In an IoT environment without basic security measures, it is easy for attackers to generate large quantities of random traffic.

While IoT application protocols such as MQTT can have security support from the transport layer (TLS), there are no mechanisms to protect IoT devices from denial of service (DoS).

Intrusion detection systems older such as Zeek and Snort [9], even if they implement network-level security solutions, do not work well in IoT environments. This fact is due to they have signature-based approaches and this makes it difficult for them, for adapting to the relatively new and growing security problems associated with IoT devices.

To address DoS attacks on an IoT system using the MQTT protocol the first steep purposed is an environment to be able to perform the attacks and capture all network traffic.

The MQTT protocol is a lightweight M2M (machine-to-machine) and a publish/subscribe protocol with a minimal packet overhead. MQTT is lighter in order to transfer data compared with other protocols running over application layer like HTTP. In MQTT architecture the broker is the responsible for managing the connections. Broker connect clients with each other, a client can be devices working like sensors, actuators, pc or smartphone applications. Therefore, clients subscribe to topics where they can publish or receive information in real time. This operation can be seen in detail in Fig. 1. Representing two client applications that can receive data from the sensor and interact with the actuator.

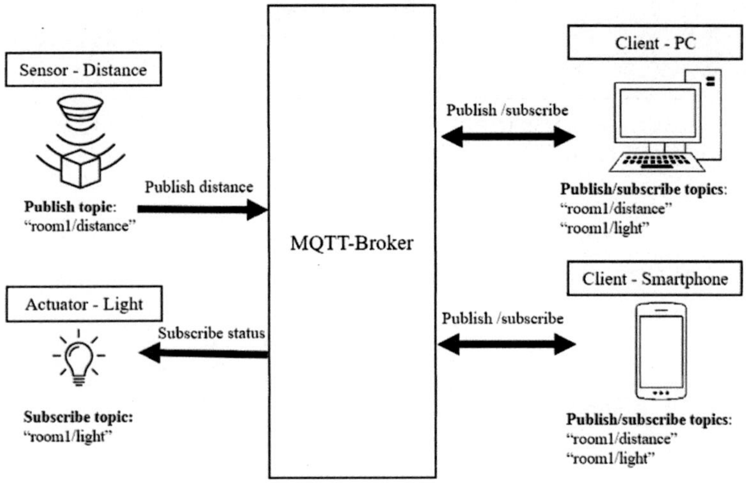

Fig. 1. MQTT environment

It is developed an MQTT environment to simulate the real traffic. A broker programmed in node.js with the Aedes library. It uses an actuator with a relay, a distance sensor, and two clients, smartphone and computer, that in addition to browsing through the Internet and interact with the IoT system. All the traffic generated in this environment is captured by a router with the OpenWRT OS installed.

Several DoS attacks are performed on the environment considering the vulnerabilities of the protocol. In a MQTT system without authentication an attacker could scan the network with search engine like Shodan through the well-known port 1883 and discover these systems. A vulnerable part of the system is the broker, which centralizes all control of the system.

The attacks are performed using the MQTT Malaria tool for send a large number of diffuse messages (of 100 characters) simulating 1000 clients sending 1000 messages per second. In such a way that the broker cannot respond to all of them, causing failure of the all IoT system. To generate the dataset of the test environment, all the traffic generated from the router is captured to obtain a pcap file. Due to the large quantity in the pcap file, it is simplified by using a tool developed for the research that does the following tasks:

- Organizes the frames of the pcap file. In the Dos attacks, numerous frames are produced in a short period of time. The capture tool (tcpump inside the router with OpenWRT) overlaps several frames in the same timestamp. So, it is necessary to separate them if they are within the same range of the timestamp, which can provide relevant information about the attack.
- Dissects the frames by fields common to all the frames, following the example of the AWID dataset [14], which is also about 802.11, taking all the fields of the MQTT protocol. It also takes all the fields of the MQTT protocol, resulting 65 fields.
- Tag each frame taking into account the timestamp of when starts and ends each attack for classifying them as in attack or not.

The dataset resulting is a CSV file with the following information: 94.625 frames. 45.513 of them tagger like under attack traffic and 49.112 like "normal" traffic [1].

3 Soft Computing Techniques Used

The proposal consists of a first level, in which a feature extraction technique is implemented, and a second level that corresponds to the classification method. Therefore, for a better understanding, this section is divided in two subsections.

3.1 Feature Extraction Methods

Due to the fact to the wide dimensionality of the dataset used (65 fields), and with the aim of reducing the training execution time, a dimensional reduction technique is implemented. Therefore, in the present research Principal Component Analysis method is used.

Principal Component Analysis. Principal Component Analysis (PCA) is a unsupervised multivariate statistical technique introduced by Pearson [17] and commonly used for dimensional reduction. This method describes the variation of a multivariate dataset as a set of uncorrelated variables, corresponding to linear combinations of the original variables. Generally speaking, the main goal of this technique is to obtain a new set of orthogonal axes in which the data variance is maximized. This is performed through the calculation of the eigenvalues of the correlation matrix. Then, using the eigenvectors, the initial set can be linearly transformed into lower dimension space.

3.2 Classification Methods

Once the feature extraction method used has been described, the supervised classification techniques are listed.

Random Forest. Random Forest (RF) is a supervised machine learning technique frequently used in classification problems, although it can also be used in regression. This method was proposed by Breiman, and it is based on the combination of decision trees and Bootstrap Aggregation ensemble technique.

The Bootstrap Aggregation method, also known as Bagging, is the ensemble technique used to generate random subsets of the whole dataset to fit each decision tree. Bagging selection method corresponds to a replacement selection process, so the same data sample can be in more than one subset. Finally, with all the decision trees trained independently, the random forest classification output corresponds to the class obtained in most of decision trees.

Support Vector Machine. Support Vector Machines (SVMs) are a set of machine learning algorithms introduced by Cortes and Vapnik [4]. This kind of algorithms are commonly use both in classification and regression problems. The main goal of SVMs is to find a hyperplane which maximize the minimum distance between the hyperplane and the samples of each class closest to the calculated hyperplane (as know as margin).

The above definition of SVMs assumes that the classes can be linearly separated. However, most real datasets cannot be linearly separated. In these cases, SVMs algorithms implement data transformations, $\langle x_i, x_j \rangle \rightarrow \langle \phi(x_i), \phi(x_j) \rangle$, to map it into a higher dimensional space in where the dataset can be linearly separated. The transformation function $\phi(x)$ is determined by the selection of a certain kernel function.

Fisher Linear Discriminant. Fisher Linear Discriminant Analysis, also know as Linear Discriminant Analysis (LDA), is a method developed by R.A. Fisher, widely used in statistics and in machine learning classification problems.

General speaking, the main goal of this method is to find the ideal hyperplane that allows the data to be projected on it, and separating data classes as much as possible. To achieve this, LDA looks for the hyperplane in which the classes have the minimum variance between their data and in which the means of each class are as far as possible. The optimization problem is based on maximizing objective function $J(\Theta)$ defined by Eq. 1.

$$J(\theta) = \frac{(\mu_1 - \mu_2)^2}{\hat{s}_1^2 + \hat{s}_2^2} \tag{1}$$

where μ_1 and μ_2 corresponds to mean value of class 1 and 2, and $\hat{s}_1$ and $\hat{s}_2$ define within-class variance 1 and 2.

Naive Bayes. Naive Bayes, or Naive Bayesian (NB), are simple machine learning techniques based on Bayes' statistical theorem and commonly used in classification problems. In addition, these methods assume that data attributes are conditionally independent given the class. In general, this assumption is too strong, nonetheless Naive Bayes performance achieves competitive results with high computational efficiency.

There are different types of NB algorithms. In this research, Gaussian and Bernoulli Naive Bayes have been implemented.

- ***Gaussian Naive Bayes:*** In Gaussian NB, the numeric attributes values are normally distributed and are represented relative to the mean and standard deviation. This algorithm obtains the probability of the features by means of Eq. 2 (Gaussian equation).

$$P(x_i \mid c) = \frac{1}{\sqrt{2\pi\sigma_c^2}} exp\left(-\frac{(x_i - \mu_c)^2}{2\sigma_c^2}\right) \quad (2)$$

  where σ corresponds to the standard deviation and μ to the mean value.
- ***Bernoulli Naive Bayes:*** Bernoulli NB assumes that each feature corresponds to binary values. The probability is calculated through Eq. 3.

$$P(x_i \mid c) = P(x_i \mid c)b_i + (1 - b_i)(1 - p(x_i \mid c)) \quad (3)$$

  This algorithm requires that all data features were binary, if any feature includes any other kind of data, a binarization process is executed.

4 Experiments and Results

The present section describe the setup of the experiments and the results obtained. For a better understanding and to simplify the present section, it has been divided in two different subsections.

4.1 Experiments

In this subsection the setup used to execute the experiments, as well as the metrics used to evaluate and compare each of the classifiers, are described. The experiment was carried out using the Python programming language and the scikit learn machine learning library.

Data Preprocessing. Firstly, data preprocessing is mandatory. In this step, the dataset variables that were constant for all samples were discarded since they did not provide information. In addition, each categorical variable was converted to natural number by means of natural codification. Finally, the dataset was normalized using the z-score method (mean value 0 and standard deviation 1).

Feature Extraction. Once the dataset was preprocessed, the PCA method is executed to reduce dataset dimensionality. The number of principal components to be used will be determined from an initial analysis, taking into account the components with a high percentage of variance.

Classifier Setup. The different classification methods have been tested for different settings of their hyperparameters. These configurations are listed next:

- RF: models were created for different numbers of decision trees. In this research the performance was tested for models with 10, 20, 30 and 40 decision trees.
- SVM: for the SVM models, it was tested for two different kernel types (linear and polynomial). In addition, different values of the regularization parameter, C, were tested. In this case, the values of C were 1, 0.1 and 0.001.
- LDA: in this case two different algorithm solver were checked (singular value decomposition, 'svd', and least squares, 'lsqr').
- NB: Bernoulli and Gaussian Naive Bayes performance were evaluated.

Classifier Evaluation. The models created were trained following a k-fold cross-validation method, using $k = 10$. Then, models of the same type are compared with each other to select the best performing configuration. Finally, the best models are compared to obtain the best performance. Finally, the best models are compared to obtain the best performance. The metrics measured were accuracy, F1-score, precision, recall, specificity, and the Area Under the receiving operating Curve (AUC). To compare the models and determinate the best classifier AUC is used since it is insensitive to changes in class distribution. Also, to compare the models, Kruskal-Wallis variance analysis and Tukey comparison method is used if differences between the models were determined.

4.2 Results

After explaining the methods used and the experiments carried out, the results are presented. Firstly, an initial PCA analysis was proposed to determine the number of components to be considered. Thus, the percentage of variance explained by a large number of components was analyzed. Figure 2 shows the result of the initial PCA analysis.

Taking into account the results shown in Fig. 2, and considering the components with a percentage of variance greater than 5%, a dimensionality reduction of 5 components was considered.

Once the results of the initial PCA analysis are shown, the metrics obtained with each model are displayed. Table 1 shows results obtained for all random forest configurations. In this case, the RF method obtains good results for any configuration with AUC percentages higher than 99%. The best results are obtained with a RF of 30 trees.

Table 2 shows the results obtained with SVMs. In this case the best AUC metrics are obtained when the method is used with a smooth regularization (regularization is inversely proportional to C). In particular, the best SVM results are obtained with a polynomial kernel and $C = 1$.

Finally, Table 3 shows the results for LDA, whereas Table 4 shows the metrics obtained for Gaussian and Bernoulli NB. With both techniques, the AUC metrics

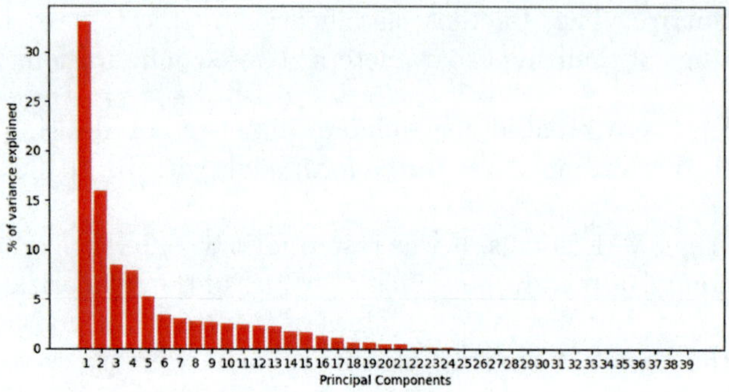

Fig. 2. Initial PCA

Table 1. Random forest results.

Random Forest						
Nº of trees	**Accuracy**	**AUC**	**Recall**	**Precision**	**f1_score**	**Specificity**
10	0,996583	0,992965	0,986371	0,998468	0,992381	0,999559
20	0,996618	0,993067	0,986598	0,998391	0,992458	0,999537
30	**0,99672**	0,993241	**0,986901**	**0,998546**	**0,992687**	**0,999581**
40	0,996549	0,993023	0,986598	0,998086	0,992308	0,999449

Table 2. Support vector machines results.

Support Vector Machines							
Kernel	**C**	**Accuracy**	**AUC**	**Recall**	**Precision**	**f1_score**	**Specificity**
linear	1.0	0,911854	0,840653	0,710911	0,887924	0,774275	0,970396
	0.1	0,894788	0,788794	0,595656	0,918498	0,70991	0,981933
	0.001	0,88242	0,74642	0,498598	0,962007	0,656582	0,994242
polynomial	**1.0**	**0,949282**	0,907317	**0,830848**	**0,937452**	**0,880862**	**0,983786**
	0.1	0,889151	0,755274	0,511319	0,994835	0,675274	0,999228
	0.001	0,886161	0,748675	0,498144	0,994599	0,663632	0,999206

Table 3. Fisher linear discriminant results.

Fisher Linear Discriminant						
Solver	**Accuracy**	**AUC**	**Recall**	**Precision**	**F1_score**	**Specificity**
svd	0,881874	0,745209	0,496175	0,961834	0,654438	0,994242
lsqr	0,881874	0,745209	0,496175	0,961834	0,654438	0,994242

obtained are lower than those of RF and SVM. In the case of LDA, changes in the solver do not affect the value of the metric and the AUC reaches 74.5%. On the other hand, the best results in NB are achieved by the Bernoulli model with an AUC of 78.1%.

Table 4. Naive bayes results.

Naive Bayes						
Type	**Accuracy**	**AUC**	**Recall**	**Precision**	**F1_score**	**Specificity**
Gaussian	0,881788	0,75111	0,512985	0,932901	0,661751	0,989235
Bernoulli	**0,887152**	0,781918	**0,590152**	**0,876322**	**0,698972**	**0,973683**

In conclusion, of the methods proposed in this research, the most suitable method for detecting DoS attacks in the MQTT environment is the random forest method, which achieves an AUC metric greater than 99%.

5 Conclusions and Futures Works

In the present research four of the main supervised machine learning techniques (RF, SVM, LDA and NB), combined with PCA, have been tested to detect DoS attacks in MQTT networks. The results obtained show good functionality with RF technique, which obtained a value of AUC and the other measured metrics higher than 99%. Also, SVM with polynomial kernel shows good results.

Due to the good performance of supervised machine learning algorithms combined with PCA dimensional reduction technique, other techniques based on different idea would be applied [3,8]. For instance, hybrid models [5–7,12] or one-class techniques [10,11,15] can be considered to compare with supervised methods. On the other hand, the use of other dimensional reduction techniques could be considered to compare the results obtained.

Acknowledgements. Spanish National Cybersecurity Institute (INCIBE) and developed Research Institute of Applied Sciences in Cybersecurity (RIASC).

CITIC, as a Research Center of the University System of Galicia, is funded by Consellería de Educación, Universidade e Formación Profesional of the Xunta de Galicia through the European Regional Development Fund (ERDF) and the Secretaría Xeral de Universidades (Ref. ED431G 2019/01).

References

1. MQTT Dataset LE-229-18 (2019). https://joseaveleira.es/dataset
2. Aversano, L., Bernardi, M.L., Cimitile, M., Pecori, R.: A systematic review on Deep Learning approaches for IoT security (2021)
3. Casado-Vara, R., Sittón-Candanedo, I., la Prieta, F.D., Rodríguez, S., Calvo-Rolle, J.L., Venayagamoorthy, G.K., Vega, P., Prieto, J.: Edge computing and adaptive fault-tolerant tracking control algorithm for smart buildings: a case study. Cybernet. Syst. **51**(7), 685–697 (2020)
4. Cortes, C., Vapnik, V.: Support-vector networks. Mach. Learn. **20**(3), 273–297 (1995)
5. Fernandez-Serantes, L.A., Casteleiro-Roca, J.L., Berger, H., Calvo-Rolle, J.L.: Hybrid intelligent system for a synchronous rectifier converter control and soft switching ensurement. Eng. Sci. Technol. Int. J. 101189 (2022)

6. Fernandez-Serantes, L.A., Casteleiro-Roca, J.L., Calvo-Rolle, J.L.: Hybrid intelligent system for a half-bridge converter control and soft switching ensurement. Revista Iberoamericana de Automática e Informática industrial (2022)
7. García-Ordás, M.T., Alaiz-Moretón, H., Casteleiro-Roca, J.L., Jove, E., Benítez-Andrades, J.A., García-Rodríguez, I., Quintián, H., Calvo-Rolle, J.L.: Clustering techniques selection for a hybrid regression model: a case study based on a solar thermal system. Cybernet. Syst. 0(0), 1–20 (2022)
8. Gonzalez-Cava, J.M., Arnay, R., Mendez-Perez, J.A., León, A., Martín, M., Reboso, J.A., Jove-Perez, E., Calvo-Rolle, J.L.: Machine learning techniques for computer-based decision systems in the operating theatre: application to analgesia delivery. Log. J. IGPL **29**(2), 236–250 (2020)
9. Hamza, A., Gharakheili, H.H., Benson, T.A., Sivaraman, V.: Detecting volumetric attacks on IoT devices via SDN-based monitoring of MUD activity. In: SOSR 2019 - Proceedings of the 2019 ACM Symposium on SDN Research, pp. 36–48. Association for Computing Machinery, Inc (2019)
10. Jove, E., Casteleiro-Roca, J.L., Casado-Vara, R., Quintián, H., Pérez, J.A.M., Mohamad, M.S., Calvo-Rolle, J.L.: Comparative study of one-class based anomaly detection techniques for a bicomponent mixing machine monitoring. Cybernet. Syst. **51**(7), 649–667 (2020)
11. Jove, E., Casteleiro-Roca, J.L., Quintián, H., Méndez-Pérez, J.A., Calvo-Rolle, J.L.: A new method for anomaly detection based on non-convex boundaries with random two-dimensional projections. Inf. Fusion **65**, 50–57 (2021)
12. Jove, E., Gonzalez-Cava, J.M., Casteleiro-Roca, J.L., Quintián, H., Méndez Pérez, J.A., Vega Vega, R., Zayas-Gato, F., de Cos Juez, F.J., León, A., MartÍn, M., Reboso, J.A., Wozniak, M., Luis Calvo-Rolle, J.: Hybrid intelligent model to predict the remifentanil infusion rate in patients under general anesthesia. Logic J. IGPL **29**(2), 193–206 (2020)
13. Khalid, M.H., Murtaza, M., Habbal, M.: Study of security and privacy issues in internet of things. In: 2020 5th International Conference on Innovative Technologies in Intelligent Systems and Industrial Applications (CITISIA), pp. 1–5. IEEE (2020)
14. Kolias, C., Kambourakis, G., Stavrou, A., Gritzalis, S.: Intrusion detection in 802.11 networks: Empirical evaluation of threats and a public dataset. IEEE Commun. Surveys Tutor. **18**(1), 184–208 (2016)
15. Leira, A., Jove, E., Gonzalez-Cava, J.M., Casteleiro-Roca, J.L., Quintián, H., Zayas-Gato, F., Álvarez, S.T., Simic, S., Méndez-Pérez, J.A., Luis Calvo-Rolle, J.: One-class-based intelligent classifier for detecting anomalous situations during the anesthetic process. Log. J, IGPL (2020)
16. Liu, J., Kantarci, B., Adams, C.: Machine learning-driven intrusion detection for contiki-NG-based IoT networks exposed to NSL-KDD dataset. In: Proceedings of the 2nd ACM Workshop on Wireless Security and Machine Learning. ACM, New York, NY, USA (2020)
17. Pearson, K.: Liii. on lines and planes of closest fit to systems of points in space. The London, Edinburgh, and Dublin Philosophical Magaz. J. Sci. **2**(11), 559–572 (1901)
18. Wilson, D.R.: Towards effective wireless intrusion detection using AWID dataset. Theses (2021)

Automating the Implementation of Unsupervised Machine Learning Processes in Smart Cities Scenarios

Raúl López-Blanco[1(✉)], Ricardo S. Alonso[1,2], Javier Prieto[1], and Saber Trabelsi[3]

[1] BISITE Research Group, University of Salamanca, Edificio Multiusos I+D+i, Calle Espejo 2, 37007 Salamanca, Spain
{raullb,ralorin,javierp}@usal.es

[2] AIR Institute - Deep tech lab IoT Digital Innovation Hub, Salamanca, Spain
ralonso@air-institute.com

[3] Texas A&M University at Qatar, Doha, Qatar
saber.trabelsi@qatar.tamu.edu

https://bisite.usal.es, https://air-institute.com

Abstract. Climate Change has become a problem for all the inhabitants of the planet and the solutions to curb it involve knowing all the data on its causes and effects. To this end, it is essential to have mechanisms capable of reading data from different media in real time. This will make it possible to solve many of the problems that arise in areas such as medicine, Smart Cities, industry, transport, etc. Analysing raw data to provide it with semantics is essential to exploit its full potential, making it possible to manage a large number of everyday tasks. All this raw data often comes from a large number of sensors and other sources, in very different types and formats. The analysis of this data read in real time and cross-referenced with information stored in heterogeneous databases, with data from simulations or with data from digital twins is a great opportunity to combat problems such as Climate Change. This work presents a successful use case by characterising the city of Salamanca in vegetation clusters, where a decarbonisation process of a communication artery that crosses the city from north to south is being carried out. The results of this study will serve to identify the most necessary areas for action in the fight against the polluting gases that cause Climate Change.

Keywords: Climate change · Smart cities · Unsupervised machine learning · Artificially intelligent city · Vegetation index

1 Introduction

Today, Climate Change is one of humanity's main concerns. Data from Eurobarometer 2020 [10] confirms this, indicating that 91% of Europeans are concerned about Climate Change and 83% believe that policies are needed to combat this problem.

J. M. Machado et al. (Eds.): DCAI 2022, LNNS 585, pp. 71–80, 2023.
https://doi.org/10.1007/978-3-031-23210-7_7

This problem affects not only humans but also the rest of the species on planet Earth and society is determined to reverse the causes of this problem. The effects of this problem [14], are already being felt today [30] and are likely to continue worsening in the coming years [4].

This challenge must be addressed through local strategies because Climate Change does not affect all territories in the same way. There are more and more regional strategies applied to this problem [5] due to the information that can be obtained from cities and/or territories is increasing. New developments in the field of computer science allow for the application of programmatic processes to large volumes of data [12] [13]. Looking for the application of processing, analysis and Artificial Intelligence techniques [25] on data, platforms such as the one described in this work [20].

So far, it has not been possible to achieve all the objectives due to overly ambitious and global proposals, which see the smart city as a single and indivisible entity [24]. Against this idea appeared the platform used in the scenario, which advocates a flexible and modular architecture that allows adaptation to different use cases.

The main objective of this use case is to propose a practical and holistic approach to climate change adaptation and the enhancement of biodiversity and ecosystem services. Its interest lies in its great potential for replication in cities and/or with heritage centres across Europe.

The rest of this work is described as folows: in Sect. 2, where reference is made to related work that has served to inspire it and define its scope of study. After this, Sect. 3 presents the functionalities of the platform used in the success case. Then, in Sect. 4, the process of applying the platform to the use case is presented. Finally, in the last Sect. 5, conclusions and lines of future work are presented.

2 Related Works

In recent years, the studies carried out on Climate Change have undergone a revolution thanks to the appearance of Big Data as a means of handling the enormous set of data needed for this type of studies [22]. However, this is a complex change, since the Earth is a changing and complex system and there are a large number of variables that influence this process [18]. This is why, until now, due to the novelty of the technology and its magnitude, it has not yet exploited its full potential.

Added to this novel technology is also the Internet of Things (IoT) and the importance of Smart Cities and all the data that can be collected from them [23] especially in the last decade. Data has become an essential part of Smart Cities and takes an important role in veracity, estimation and decision making processes [3]. IoT has been applied in recent times in a number of solutions such as Smart Agriculture [2], Smart Consumption [20] and even in network architectures [1]. Looking at most of the environments where IoT technology is applied, from Industrial IoT [6] to maritime [26], all of these sectors ultimately relate to Smart Cities [11], which are evolving thanks to the application of this technology along with others such as blockchain [15] and other paradigms such as edge computing [27].

Along with Big Data, Artificial Intelligence and Machine Learning are other great allies when carrying out certain studies and predictions on Climate Change and have been used both at regional level [16], and at continental level [7], and is that this type of algorithms are able to correlate large amounts of data and get to make predictions about future scenarios in times that until a few years ago would have been unthinkable.

With the solution provided, the aim is to carry out a data analysis procedure that integrates the three aspects of this problem, which are the collection of data, since they come from heterogeneous sources, their storage in a consistent Big Data and the application of Machine Learning techniques and algorithms. Some systematic reviews such as the one presented in [21] point out that there are problems regarding the lack of definition of smart city standards which causes gaps in the architecture, framework and management standards for data that need to be solved. It is also noted in [21] that there are certain challenges that must be taken on by the technologies used for data dissemination used in these Smart Cities such as scalability and distributed access to data.

These problems that other systems have do not appear in the platform applied in the use case to be described below, since it is fully scalable dynamically and faces distributed access to data with a layered infrastructure that transports the data from ingestion to export through the phases of management, analysis and visualization without leaving the same [20].

3 Framework for Intelligent Data

The creation of Smart Cities requires a number of mature components that respect security and privacy standards. Data acquisition, processing and dissemination processes must also apply principles such as interoperability, consistency and reuse [21]. The need for these standards is due to the fact that the concept of Smart Cities is fairly recent and the framework for these trends is not yet too well defined [9].

Therefore, many researchers have proposed new architectures to collect all the data generated by these platforms by reusing functionalities from other applications to be able to process the huge amounts of data [8] collected by IoT devices that are placed in Smart Cities whose heterogeneity poses a challenge when generating standardized connectors [19].

Among these data management platforms is deepint.net, a platform that presents a solution to the management needs of cities, regions and territories. Some of its most important modules, which will be used in Sect. 4, are presented below.

3.1 Data Ingestion Layer

This section explains the data ingestion layer of the platform. It is responsible for handling the data that is added to it. All data is stored in entities called sources, which are the grouping unit of information that is used as the basis for the rest of the processing. Data sources can come from different places, e.g. databases, API's, URL's, files and endpoints (see Fig. 1).

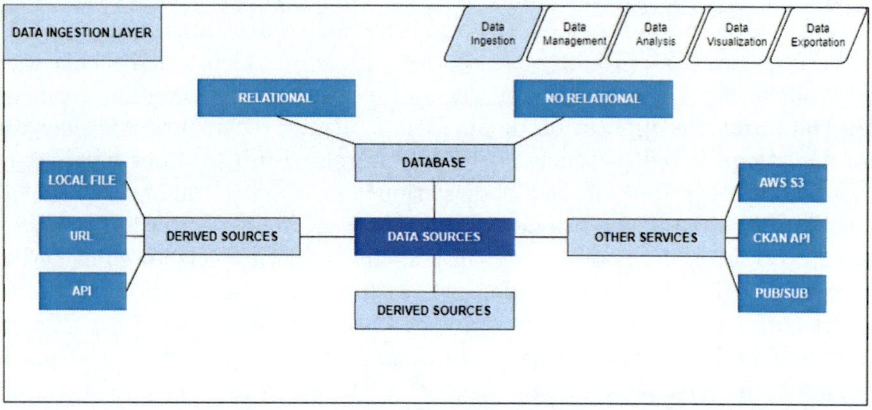

Fig. 1. Platform data ingestion layer [12].

3.2 Data Analysis Layer

Another important feature of the platform lies in its analysis layer, which is the point at which the data is acted upon (see Fig. 2). This layer allows different types of algorithms related to Machine Learning to be applied on the data. Among the applicable algorithms, there are several to solve problems of regression, classification, clustering, rule discovery, dimension reduction or text classification.

3.3 Data Visualization Layer

Another strong point of the platform is the possibility it offers to create attractive graphs for the user, allowing it to understand the information obtained from the sources in different ways (see Fig. 3). Among the graphical possibilities offered, it is possible to change colours, titles, series, filters within each graph, which can be of many types including maps, bar charts, line charts, date charts, pie charts, KPIs, heat maps, among others.

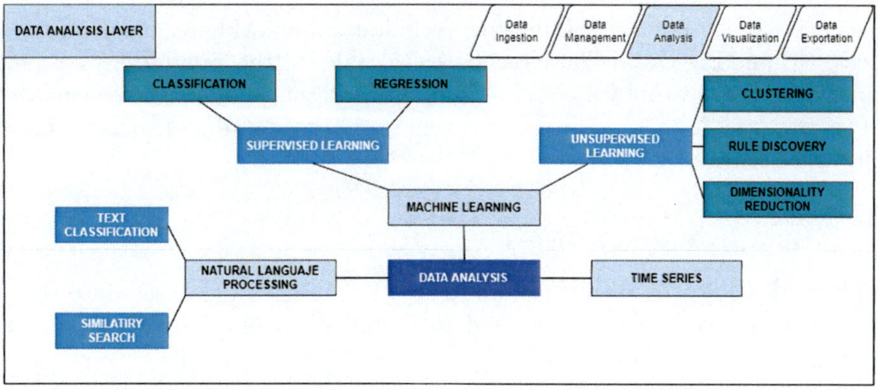

Fig. 2. Platform data analysis layer [12].

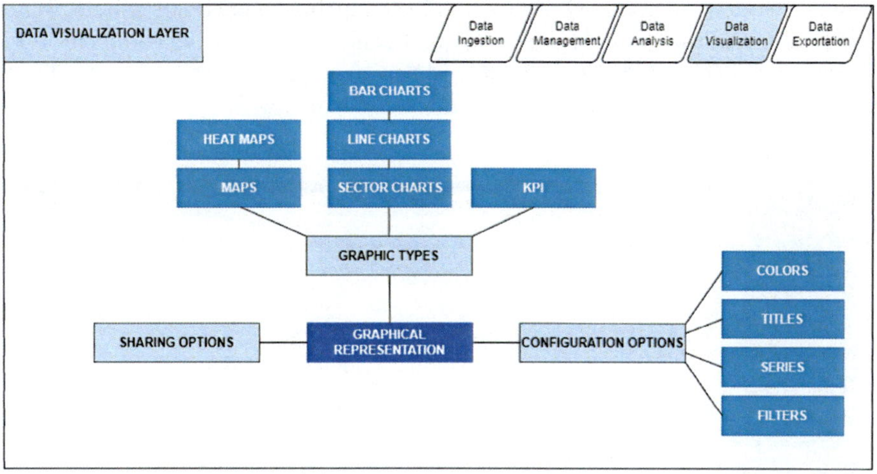

Fig. 3. Platform data visualization layer [12].

4 Decarbonization Proposal: A Case Use

After studying the characteristics of the platform, this section will explain how it has been applied to a real use case for the monitoring of a series of parameters that will help the decarbonization of one of the communication arteries of the heritage city of Salamanca.

The purpose of this use case is to obtain information for the application of actions in the field of Green Infrastructure for a greater presence and diversification of species in the environment, actions to reduce the social vulnerability of the city of Salamanca to climate change supported by the European directives [31] and in search of the reduction of the urban heat island effect [17].

All these actions and objectives are supported by an intensive use of ICT that will be integrated into sensors, thanks to IoT technology to collect data (see Sect. 3.1), Artificial Intelligence techniques applicable on the platform (see Sect. 3.2) and Big Data. These technologies could lay the foundations for a future smart city that starts by seeking to minimise carbon emissions. The visualisation of results (see Sect. 3.3) will help to achieve the promotion of social values and cultural heritage.

4.1 Data Ingestion Process

This section looks at the procedure followed to bring the datasets into the platform and how they have been used to perform the analysis. The first dataset corresponds to the air quality stations of the Junta de Castilla y León (Government of Castile and León). In this case it is a source that is ingested from the open data portal, obtaining sources for each air component. From these sources, data is collected for certain components of the city's air such as ozone (O_3), carbon monoxide (CO), nitrogen dioxide (NO_2), airborne particles smaller than 2.5 and 10 mm and the geographical location of the stations. After going through the filtering processes, visualisations have been made as shown in Fig. 4, obtaining an evolution of how the air components vary over time.

Fig. 4. Air quality stations graph.

Next, data from the European Copernicus mission [28] concerning the normalised difference vegetation index (NDVI) [29] have been taken and joined to the data from the positions of the air quality stations of the Junta de Castilla y León.

4.2 Data Analysis with Clustering and Classification Processes

This subsection explains the unsupervised Machine Learning processes followed to obtain results that lead to favourable conclusions with the data obtained in Sect. 4.1.

To work with this data, an unsupervised learning algorithm has been selected that divides a data set into clusters whose internal properties are "similar". The number of clusters k is previously defined according to mathematical criteria of "closeness" (elbow method), so that each observation (or new observations) is assigned to the cluster whose distance is smaller. This assignment process is repeated until the centroids of each group do not move or their movement is below a threshold distance. The elbow method for these data determines that the best value of k for these data is 4. This whole process is governed by the equation Eq. 1

$$\min_{\mathbf{S}} E\left(\boldsymbol{\mu}_i\right) = \min_{\mathbf{S}} \sum_{i=1}^{k} \sum_{\mathbf{x}_j \in S_i} \left\|\mathbf{x}_j - \boldsymbol{\mu}_i\right\|^2 \tag{1}$$

The objects are represented by real vectors of b dimensions $(\mathbf{x}_1, \mathbf{x}_2, \ldots, \mathbf{x}_n)$. The k-means algorithm constructs **k** groups where the sum of distances of the objects, within each group $\mathbf{S} = \{S_1, S_2, \ldots, S_k\}$, to their centroid is minimized. We define **S** as the data sets whose elements are the objects $\mathbf{x}_j$ that are vectors. We have k groups with their centroids which are updated by the condition of the function $E\left(\boldsymbol{\mu}_i\right)$ which, for the quadratic function is Eq. 2

$$\frac{\partial E}{\partial \boldsymbol{\mu}_i} = 0 \Longrightarrow \boldsymbol{\mu}_i^{(t+1)} = \frac{1}{\left|S_i^{(t)}\right|} \sum_{\mathbf{x}_j \in S_i^{(t)}} \mathbf{x}_j \tag{2}$$

This algorithm works correctly for continuous variables, in this case we are using geographical data and the NDVI vegetation index to find the groups that divide the city of Salamanca into similar zones. The implementation of the k-means algorithm also allows to parameterise the weight of the variables involved and thus obtain different groupings based on this configuration of relative weights. Figure 5 shows the areas grouped according to the different weights of the geographical variables (latitude and longitude) and the NDVI variable. The graph on the left shows the geographical points, the cluster to which each of them belongs and the location of the stations (marked with an X). The graph on the right shows some of these points on the map of Salamanca and the location of the stations (in grey).

5 Conclusions and Future Work Lines

This section will present the conclusions drawn from the analysis of the platform and future lines of work. Thanks to the results seen in the work, it has been possible to confirm that the use made of the Deepint.net platform makes it an excellent manager of heterogeneous sources. The following conclusions are drawn:

- The exposed platform applied for the use case solves the problem of heterogeneity of sources in the Smart Cities environment, allowing to gather different data sources in the same platform.

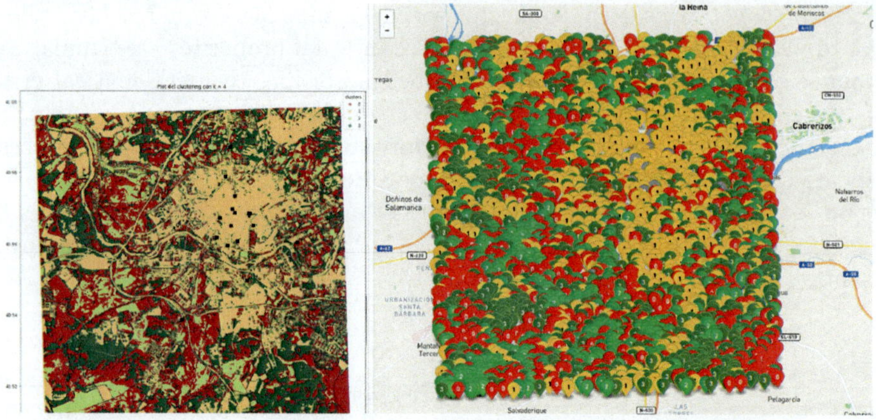

Fig. 5. Clustering in the heritage city of Salamanca with NDVI data.

- The tools provided by the platform allow to perform management and analysis functionalities in a uniform way on the collected data.
- The platform's visualizations make it possible to represent the data in a wide variety of ways to extract all the knowledge from the data, facilitating predictions and future actions.

The future of Smart Cities is still under construction, which is why some interesting avenues remain open from this work, including the following:

- Use and application of the platform in other use cases required for Smart Cities, which will allow demonstrating the versatility of the platform.
- Follow-up of the use case explained to see how the models behave with the new data taken by the IoT devices.

Acknowledgements. This research has been partially supported by the project "Intelligent and sustainable mobility supported by multi-agent systems and edge computing (InEDGEMobility): Towards Sustainable Intelligent Mobility: Blockchain -based framework for IoT Security", Reference: RTI2018-095390-B-C32, financed by the Spanish Ministry of Science, Innovation and Universities (MCIU), the State Research Agency (AEI) and the European Regional Development Fund (FEDER).

References

1. Alonso, R.S., Prieto, J., de La Prieta, F., Rodríguez-González, S., Corchado, J.M.: A review on deep reinforcement learning for the management of SDN and NFV in edge-IoT. In: 2021 IEEE Globecom Workshops (GC Wkshps), pp. 1–6. IEEE (2021)
2. Alonso, R.S., Sittón-Candanedo, I., Casado-Vara, R., Prieto, J., Corchado, J.M.: Deep reinforcement learning for the management of software-defined networks in smart farming. In: 2020 International Conference on Omni-layer Intelligent Systems (COINS), pp. 1–6. IEEE (2020)

3. Assiri, F.: Methods for assessing, predicting, and improving data veracity: a survey. ADCAIJ: Adv. Distrib. Comput. Artif. Intell. J. **9**(4), 5 (2020)
4. to Bühne, H.S., Tobias, J.A., Durant, S.M., Pettorelli, N.: Improving predictions of climate change–land use change interactions. Trends Ecol. Evolut. **36**(1), 29–38 (2021)
5. Bushnell, J., Peterman, C., Wolfram, C.: Local solutions to global problems: climate change policies and regulatory jurisdiction. Rev. Environ. Econom. Pol. (2020)
6. Campero-Jurado, I., Márquez-Sánchez, S., Quintanar-Gómez, J., Rodríguez, S., Corchado, J.M.: Smart helmet 5.0 for industrial internet of things using artificial intelligence. Sensors **20**(21), 6241 (2020)
7. Carvalho, M., Melo-Gonçalves, P., Teixeira, J., Rocha, A.: Regionalization of Europe based on a k-means cluster analysis of the climate change of temperatures and precipitation. Phys. Chem. Earth, Parts A/B/C **94**, 22–28 (2016)
8. Chamoso, P., González-Briones, A., De La Prieta, F., Venyagamoorthy, G.K., Corchado, J.M.: Smart city as a distributed platform: toward a system for citizen-oriented management. Comput. Commun. **152**, 323–332 (2020)
9. Chamoso, P., González-Briones, A., Rodríguez, S., Corchado, J.M.: Tendencies of technologies and platforms in smart cities: a state-of-the-art review. Wireless Commun. Mob. Comput.**2018** (2018)
10. Commission, E.: Attitudes of Europeans towards the environment (2020). https://europa.eu/eurobarometer/surveys/detail/2257
11. Corchado, J.M.: Blockchain and its applications on edge computing, industry 4.0, iot and smart cities. Dieleman, S (2014)
12. Corchado, J.M., Chamoso, P., Hernández, G., Gutierrez, A.S.R., Camacho, A.R., González-Briones, A., Pinto-Santos, F., Goyenechea, E., García-Retuerta, D., Alonso-Miguel, M., et al.: Deepint. net: a rapid deployment platform for smart territories. Sensors **21**(1), 236 (2021)
13. Corchado, J.M., Pinto-Santos, F., Aghmou, O., Trabelsi, S.: Intelligent development of smart cities: Deepint. net case studies. In: Sustainable Smart Cities and Territories International Conference, pp. 211–225. Springer (2021)
14. Corchado, J.M.: Technologies for sustainable consumption - researchgate.net (Apr 2021). https://www.researchgate.net/profile/Juan_Rodriguez331/publication/353755163_Technologies_for_sustainable_consumption/links/610ea9491e95fe241abaae5e/Technologies-for-sustainable-consumption.pdf
15. Corchado Rodríguez, J.M., et al.: Deeptech–ai-iot in smart cities (2021)
16. Corte-Real, J., Qian, B., Xu, H.: Regional climate change in Portugal: precipitation variability associated with large-scale atmospheric circulation. Int. J. Climatol. J. Roy. Meteorolog. Soc. **18**(6), 619–635 (1998)
17. Deilami, K., Kamruzzaman, M., Liu, Y.: Urban heat island effect: A systematic review of spatio-temporal factors, data, methods, and mitigation measures. Int. J. Appl. Earth Observat. Geoinf. **67**, 30–42 (2018)
18. Faghmous, J.H., Kumar, V.: A big data guide to understanding climate change: the case for theory-guided data science. Big data **2**(3), 155–163 (2014)
19. Fan, T., Chen, Y.: A scheme of data management in the internet of things. In: 2010 2nd IEEE International Conference on Network Infrastructure and Digital Content, pp. 110–114. IEEE (2010)
20. Garcia-Retuerta, D., Chamoso, P., Hernández, G., Guzmán, A.S.R., Yigitcanlar, T., Corchado, J.M.: An efficient management platform for developing smart cities: Solution for real-time and future crowd detection. Electronics **10**(7), 765 (2021)

21. Gharaibeh, A., Salahuddin, M.A., Hussini, S.J., Khreishah, A., Khalil, I., Guizani, M., Al-Fuqaha, A.: Smart cities: A survey on data management, security, and enabling technologies. IEEE Communications Surveys & Tutorials **19**(4), 2456–2501 (2017)
22. Hassani, H., Huang, X., Silva, E.: Big data and climate change. Big Data Cognit. Comput. **3**(1), 12 (2019)
23. Heijmeijer, A.V.H., Alves, G.V.: Development of a middleware between sumo simulation tool and Jacamo framework. ADCAIJ: Adv. Distrib. Comput. Artif. Intell. J. **7**(2), 5–15 (2018)
24. Kitchin, R.: The promise and peril of smart cities. Comput. Law: The J. Soc. Comput. Law **26**(2) (2015)
25. Milojevic-Dupont, N., Creutzig, F.: Machine learning for geographically differentiated climate change mitigation in urban areas. Sustainable Cities Soc. **64**, 102526 (2021)
26. Plaza-Hernández, M., Gil-González, A.B., Rodríguez-González, S., Prieto-Tejedor, J., Corchado-Rodríguez, J.M.: Integration of iot technologies in the maritime industry. In: International Symposium on Distributed Computing and Artificial Intelligence, pp. 107–115. Springer (2020)
27. Sittón-Candanedo, I., Alonso, R.S., Corchado, J.M., Rodríguez-González, S., Casado-Vara, R.: A review of edge computing reference architectures and a new global edge proposal. Future Generat. Comput. Syst. **99**, 278–294 (2019)
28. Union, E.: Copernicus (2022). https://www.copernicus.eu
29. Union, E.: Normalized difference vegetation index (2022). https://land.copernicus.eu/global/products/ndvi
30. U.S., N.O., Administration, A.: It's official: July was earth's hottest month on record (2021). https://www.noaa.gov/news/its-official-july-2021-was-earths-hottest-month-on-record
31. Zhongming, Z., Wei, L., et al.: Urban adaptation to climate change in Europe 2016-transforming cities in a changing climate (2016)

Intelligent Model Hotel Energy Demand Forecasting by Means of LSTM and GRU Neural Networks

Víctor Caínzos López[1](✉), José-Luis Casteleiro-Roca[1], Francisco Zayas Gato[1], Juan Albino Mendez Perez[2], and Jose Luis Calvo-Rolle[1]

[1] Department of Industrial Engineering, University of A Coruña. CTC, A Coruña, Spain
{victor.cainzos.lopez,jose.luis.casteleiro,f.zayas.gato,jlcalvo}@udc.es

[2] Department of Computer Science and System Engineering, University of La Laguna, Tenerife, Spain
jamendez@ull.edu.es

Abstract. The hotel business consumes a significant amount of energy, requiring effective management solutions to ensure its performance and sustainability. The increased position of hotels as prosumers plus the renewable energy technologies, complicate the design of these systems, which depends on the use of reliable predict6ions for energy load. Based on artificial neural networks (ANN), regression approaches such as Long Short-Term Memory (LSTM) and Gated Recurrent Unit networks (GRU), this research proposes an intelligent model for predicting energy demand in a hotel. Validation was performed using real hotel data likened to time-series models. The resulting forecasts were remarkable, indicating a promising potential for its usage in hotel energy management systems.

Keywords: Layer · Forecasting · Energy

1 Introduction

The tourism industry has grown to be a major player in regional development and the use of raw resources and energy, both of which have an impact on the environment. According to the World Tourism Organization (UNWTO), tourism accounts for over 10% of global GDP and employs one out of every ten people on the planet. Tourism ranks fourth in terms of export volume worldwide, after only fuels, chemicals, and food. According to the UNWTO, global tourism increased by 7% in 2018 compared to the same period in the previous year, marking the largest increase in international visitor arrivals since 2010.

The hotel business consumes the most energy, second only to transportation. The typical hotel usage is between 450 and 700 kWh/m2 per year, equating to

J. M. Machado et al. (Eds.): DCAI 2022, LNNS 585, pp. 81–90, 2023.
https://doi.org/10.1007/978-3-031-23210-7_8

more than 60% of total power use. However, these figures vary greatly based on a variety of factors, including the weather and the hotel categorization [1].

To manage to reduce consumption, efficient energy measures as well as renewable energy production technologies must be considered [2]. Many studies have focused on the analysis of energy performance in hotels, showing the difficulties in the assessment [3] and proposing models to correlate the energy demand with different hotel parameters such as the occupancy, the number of beds and the number of workers [4–6].

As a result, industrial activity might be expanded in a more sustainable and environmentally friendly manner, thereby lowering the ecological impact. Furthermore, energy demand management contributes to self-sufficiency and economic effectiveness [7].

From a methodological standpoint, different approaches have been used to predict energy consumption in hotels. This paper focuses on short-term forecasting, testing in turn different prediction horizons. In this field, plenty of techniques are proposed, although the most popular approaches vary from simple forecasting [8] to time series [9–11] and artificial intelligence methods, including machine learning models such as ANN [12,13], and support vector machines (SVM) [14,15]. Recurrent Neural Networks (RNN), are the most well-known and widely applied ANN architecture with time-series data problems [16–18]. LSTM [19] and GRU [20] are evolutions designed to address the vanishing gradient issue that might occur while training regular RNNs.

There is an increasing amount of research that collects various solutions for energy management systems. Some of them apply optimization techniques with linear programming [21] and predictive control [22], proposing strategies to control the flow of energy between various generation and storage systems, loads, and the network. Others are based on the foundation of AI techniques [23] such as fuzzy logic, to build up systems for advanced energy analysis and management as well as decision-making for the evaluation of energy-saving technologies. Or aim to develop control algorithms for the different services or activities conducted, to improve energy efficiency in hotels. Also, new advanced methods are based on the use of generation and load forecasts [24].

The study described here shows the proposal of an intelligent model to efficiently predict the energy demand, based on data graciously given by a premium 5-star hotel located on the Atlantic Ocean in the south of Tenerife, Canary Islands (Spain) (28.100 N, 15.400 W). These statistics included power usage and daily meteorological data, such as temperature, pressure, rainfalls, solar irradiance, and wind components throughout a year (from 1 November 2016 to 31 October 2017 with a sampling rate of one sample/hour).

2 Methods

Across this section, all those technical approaches employed throughout the study are described. From the feature engineering procedure to build up an adapted dataset of sequences handling the raw data, based on the window sliding technique, to the inner computation of the LSTM and GRU units whereupon the model configurations are based.

2.1 Feature Engineering

The sampling time of the meteorological data has been derivated to one sample hour, representing the rate of change and defining a relationship with the power load. Also, it is necessary for inspecting and cleanup to be made, in order to understand unit ranges and assure that the models will be passing appropriately formatted data (Table 1). Such is the case of wind direction (D) and velocity (V) which were converted to more interpretative wind vector components.

$$Wind_x = V \times \cos D \quad | \quad Wind_y = V \times \sin D \tag{1}$$

Similarly, being weather data, constructive signals can be achieved by using sine and cosine transforms, to deal with and provide valuable daily periodicity information. Data traces have been split into training and test sets with a common proportion of 80–20% and no previous shuffle, until the dataset of sequences is built, to preserve the continuity of the time series. Normalization was applied to scale features using the mean and standard deviation of the training data, as long as the models have no access to the test set.

$$\overline{x_i} = \frac{x_i - \mu_x}{\sigma_x} \quad \forall i, ..., n \tag{2}$$

As far as the window sliding technique, applied to create datasets of sequences, is concerned, based on a history of successive samples from the data which conforms to the window width, the model approaches in this article are meant to provide a set of predictions. In agreement with the aforementioned, and the aim of developing and evaluating their potential through many cases of study, different window widths and forecasting horizons were considered, arising the corresponding datasets that the models will use to train and infer knowledge to be able to aid consistent future estimations. Model configurations as well as their input and output width data sequences, determined by the window settings, are explained in more detail in Sect. 3.

Table 1. Measured variables and scope of values.

Samples	Sample-time (h)	P (MW)	T (°C)	Pr (mbar)	S (h)	R (mm)	W_x (m/s)	W_y (m/s)	D_{sin}	D_{cos}
9145	1	[449, 2875]	[16.6, 31.8]	[996.4, 1018]	[0, 13.1]	[0, 10.5]	[−7.38, 12.77]	[−11.4, 9.35]	[−1, 1]	[−1, 1]

2.2 LSTM

LSTM has been developed to overcome the vanishing gradient problem in the standard RNN by improving the gradient flow within the network [19]. This is achieved using an LSTM unit in place of the hidden layer. As shown in Fig. 1, an LSTM unit is composed of [19] [25]:

Cell State: brings information along the entire sequence and represents the memory of the network.

$$C(t) = \sigma(f(t) \odot C(t-1) + i(t)) \tag{3}$$

Forget Gate: decides what is relevant to keep from previous time steps.

$$f(t) = \sigma(x(t)U_f + h(t-1)W_f) \tag{4}$$

Input Gate: manages what information is relevant to add from the current time step.

$$i_1(t) = \sigma(x(t)U_i + h(t-1)W_i) \tag{5}$$

$$i_2(t) = tanh(x(t)U_g + h(t-1)W_g) \tag{6}$$

$$i(t) = i_1(t) \odot i_2(t) \tag{7}$$

Output Gate: computes the value of the output at current time step.

$$o(t) = \sigma(x(t)U_o + h(t-1)W_o) \tag{8}$$

$$h(t) = tanh(C_t) \odot o(t) \tag{9}$$

The operator $\odot$ refers to Hadamard product [26].

2.3 GRU

GRU networks work very similarly to LSTM, with an update and reset gate to decide what information should be passed to the output [27]. These gates can be trained to either keep valuable or remove irrelevant information from prior time steps to prediction. According to Fig 2, the computation inside a GRU unit can be summarized [28].

Reset Gate: decides how much of the information from the previous time steps can be forgotten.

$$r(t) = \sigma(x(t)U_r + h(t-1)W_r) \tag{10}$$

Update Gate: select how much of the information from the previous time steps must be saved.

$$z(t) = \sigma(x(t)U_z + h(t-1)W_z) \tag{11}$$

Memory: brings information along the entire sequence and represents the current ($\hat{h}$) and the final memory (h) of the network.

$$\hat{h}(t) = tanh(x(t)U_h + (r(t) \odot h(t-1))W_h) \tag{12}$$

$$h(t) = (1 - z(t)) \odot h(t-1) + z(t) \odot \hat{h}(t)) \tag{13}$$

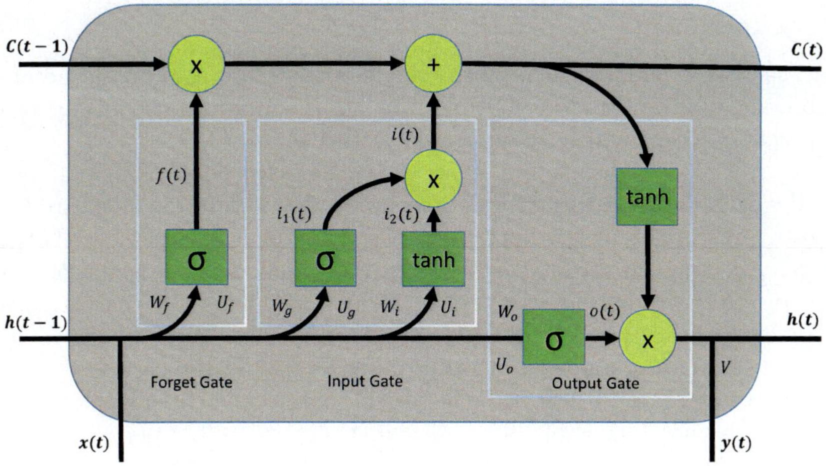

Fig. 1. A LSTM unit scheme. It is composed of the cell state, forget gate, input gate and output gate. U and W represent the weights of inputs and recurrent connections for the internal layers within each gate. σ and $tanh$ alludes to sigmoid and hyperbolic tangent activation functions each.

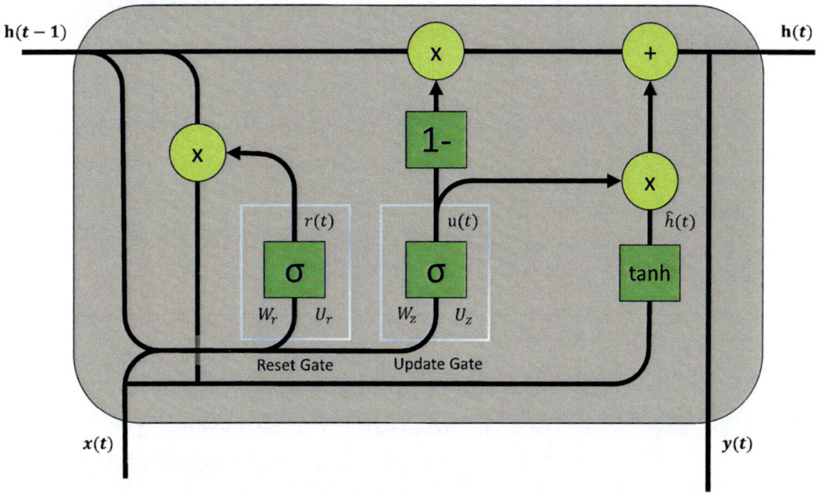

Fig. 2. A GRU unit scheme. It is composed of the reset gate, update gate and memory. U and W represents the weights of inputs and recurrent connections for the internal layers within each gate.

3 Experiments and Results

Throughout this section, all of the suggested experiments, as well as the results achieved for each simulation, are discussed.

3.1 Experiments

The suggested models will perform single-output forecasts of the hotel power load, being able to train on multiple time steps simultaneously, using two main approaches: LSTM and GRU unit architectures.

In agreement with introduced in Sect. 2.1, based on a history of successive samples of input features that mention the window size, the models are intended to yield predictions of the hotel power load through a forecasting horizon. In this line, as a nod to the forecasting potential the models may point at, different cases of study are considered, testing its performance on scenarios with both a window and horizon width of 24, 48, 72, and 96 h. Figure 3 shows the way data flow across the model configuration, according to the case of study, t may take values of 23, 47, 71, or 95 of historical and horizon time steps, taking into account the sample time of 1 h.

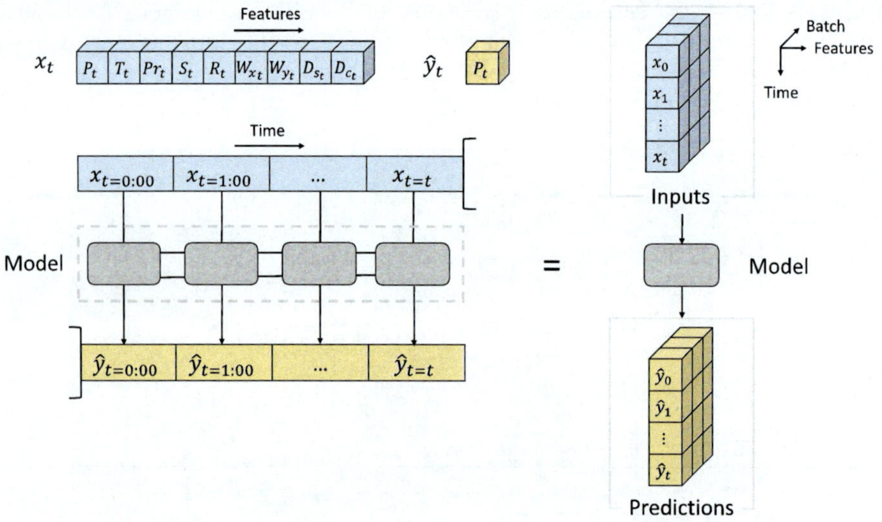

Fig. 3. Model design and data flow. Power (P), Temperature (T), Preassure (Pr), Sun (S), Rainfall (R), Wind components (Wind_x, Wind_y) and Day periodicity (Day_s, Day_c)

Models are optimized using the Cross-Validation (CV) resampling approach, to estimate their inductive reasoning in long term with the Mean Squared Error (MSE) metric as the objective function to reduce. Afterward, they are fitted on the training set and evaluated for predictions on the test data, alongside a statistical analysis fulfilled, concerning which model performs significantly better.

3.2 Results

The different simulations were accomplished considering the foregoing conditions for the window size and forecasting horizon, representing the scenarios of 24,

48, 72, and 96 h of operating range for each model configuration: LSTM and GRU unit architectures. Figure 4 draws the power load scaled predictions of the models, fitted on the full training set, throughout a range of samples that belong in the hold-out test set, representing the goodness of the future generalization.

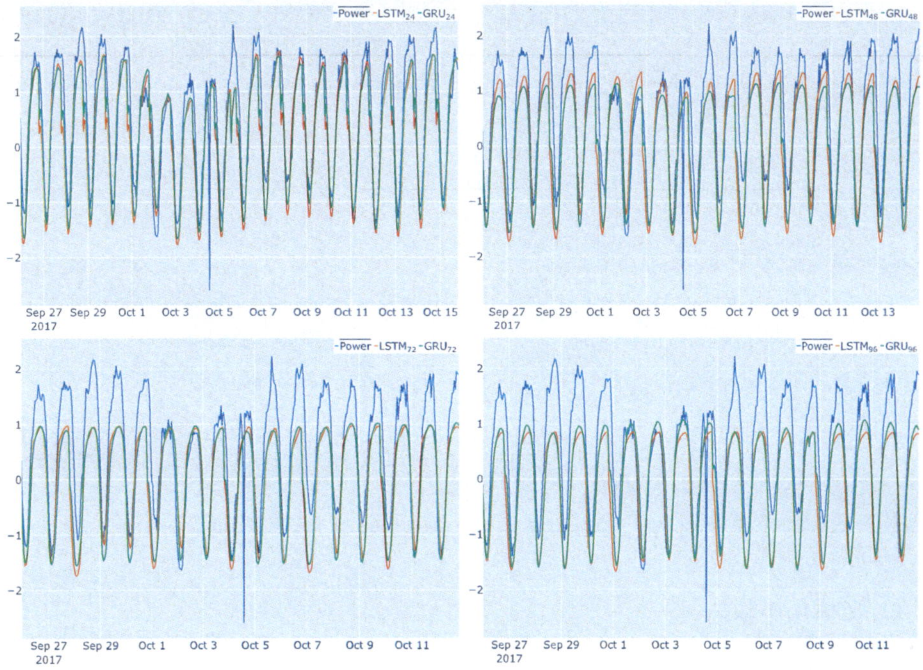

Fig. 4. LSTM and GRU scaled power load predictions for each case of study.

Likewise, variance analysis is performed to determine whether there are or are not significant variations in the score means recorded during CV resampling. Kruskal-Wallis and ANOVA methods [29] were used to determine that the null hypothesis, which asserts that all models have the same metric scores, can be rejected with a 95% confidence interval. If the null hypothesis is rejected, still another stage must be done to determine which models are distinct via multiple comparison analysis, as techniques of Tukey [30] and Holm-Bonferroni [31] which examine differences between each pair of means, regarding the appropriate multiple comparison adjustment. Figure 5 contains all of the p-values obtained from the multiple comparative analysis amid the model with the best scores and the rest. Those with p-values less than 0.05, and hence outside the 95 percent confidence interval, are highlighted, indicating that there is almost probably a significant difference between those two models.

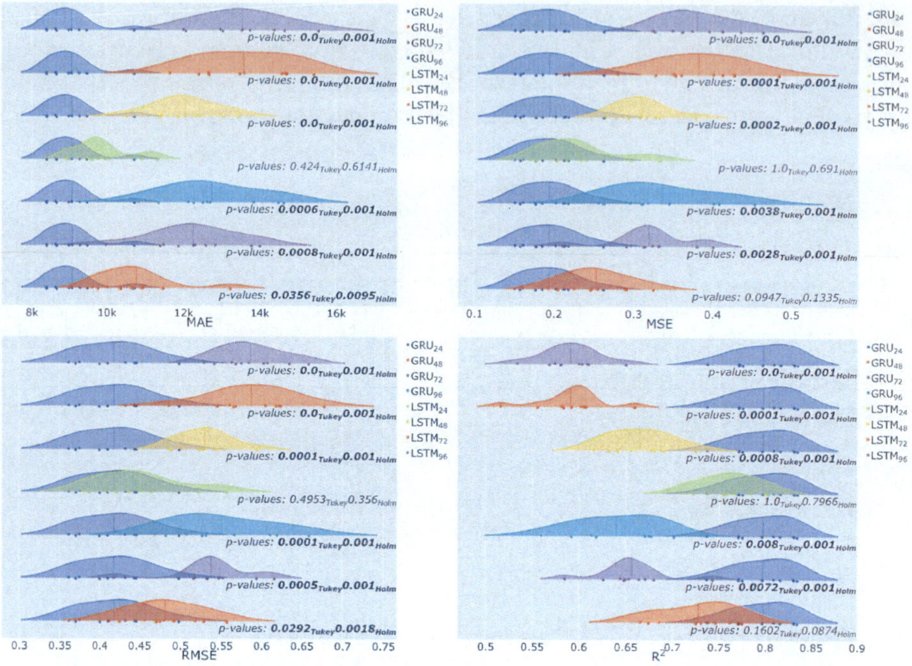

Fig. 5. Statistical comparison and p-values of the CV metric results.

4 Conclusions

Several error metrics such as Mean Absolute Error (MAE), MSE, and Root Mean Squared Error (RMSE) were used along with the statistical measure R^2 or coefficient of determination, that determines the proportion of variance in the dependent or response variable, considered as the hotel power load, that can be explained by the independent or explanatory variables, defined by the input features, thus showing the goodness of the model-data fit.

The results obtained show that the models evaluated were able to satisfactorily predict the energy load for the next 24, 48, 72, and 96 h, with a noteworthy fit to the real curve in the test set. Aforementioned notwithstanding, the model GRU_{24} appears to consistently arise the best scores. Further, gaining support with the statistical analysis confirming unanimously, for Tukey and Holm-Bonferroni methods, significant differences with almost all of the other models, except $LSTM_{24}$ at all and GRU_{48} spontaneously for MSE and R^2 metrics.

In agreement with the foregoing suggestions, the model GRU_{24} seems to be a potential tool to cope with the prediction of power demand in hotels.

Acknowledgements. CITIC, as a Research Center of the University System of Galicia, is funded by Consellería de Educación, Universidade e Formación Profesional of the Xunta de Galicia through the European Regional Development Fund (ERDF) and the Secretaría Xeral de Universidades (Ref. ED431G 2019/01).

References

1. Pieri, S.P., Tzouvadakis, I., Santamouris, M.: Identifying energy consumption patterns in the Attica hotel sector using cluster analysis techniques with the aim of reducing hotels' CO_2 footprint. Energy Build. **94**, 252–262 (2015)
2. Dalton, G.J., Lockington, D.A., Baldock, T.E.: Feasibility analysis of renewable energy supply options for a grid-connected large hotel. Renew. Energy. **34**, 955–964 (2009)
3. Deng, S.M., Burnett, J.: Study of energy performance of hotel buildings in Hong Kong. Energy Build. **31**, 7–12 (2000)
4. Papamarcou, M., Kalogirou, S.: Financial appraisal of a combined heat and power system for a hotel in Cyprus. Energy Convers. Manag. **42**, 689–708 (2001)
5. Priyadarsini, R., Xuchao, W., Eang, L.S.: A study on energy performance of hotel buildings in Singapore. Energy Build. **41**, 1319–1324 (2009)
6. Cabello Eras, J., Sousa Santos, V., Sagastume Gutiérrez, A., Guerra Plasencia, M., Haeseldonckx, D., Vandecasteele, C.: Tools to improve forecasting and control of the electricity consumption in hotels. J. Clean. Prod. **137**, 803–812 (2016)
7. Hilton Worldwide. Energy. 2018
8. Some simple forecasting methods. OTexts, March 2018
9. Atique, S., Noureen, S., Roy, V., Subburaj, V., Bayne, S., Macfie, J.: Forecasting of total daily solar energy generation using ARIMA: a case study. In: IEEE 9th Annual Computing and Communication Workshop and Conference (CCWC), Las Vegas, NV, USA, 7–9, pp. 114–119 (2019)
10. Mat Daut, M.A., Hassan, M.Y., Abdullah, H., Rahman, H.A., Abdullah, M.P., Hussin, F.: Building electrical energy consumption forecasting analysis using conventional and artificial intelligence methods: a review. Renew. Sustain. Energy Rev. **70**, 1108–1118 (2017)
11. Nguyen, H., Hansen, C.K.: Short-term electricity load forecasting with time series analysis. In: IEEE International Conference on Prognostics and Health Management (ICPHM), Dallas, TX, USA, 19–21, pp. 214–221 (2017)
12. Zúñiga, K.V., Castilla, I., Aguilar, R.M.: Using fuzzy logic to model the behavior of residential electrical utility customers. Appl. Energy. **115**, 384–393 (2014)
13. Abreu, T., Alves, U.N., Minussi, C.R., Lotufo, A.D.P., Lopes, M.L.M.: Residential electric load curve profile based on fuzzy systems. In: IEEE PES Innovative Smart Grid Technologies Latin America (ISGT LATAM), Montevideo, Uruguay, 5–7, pp. 591–596 (2015)
14. Chen, Y., Tan, H.: Short-term prediction of electric demand in building sector via hybrid support vector regression. Appl. Energy. **204**, 1363–1374 (2017)
15. Wasseem Ahmad, M., Mourad, A., Rezgui, Y., Mourshed, M.: Deep highway networks and tree-based building energy consumption. Energies **11**, 3408 (2019)
16. Graves, A., Liwicki, M., Fernandez, S., Bertolami, R., Bunke, H., Schmidhuber, J.: A novel connectionist system for improved unconstrained handwriting recognition. IEEE Trans. Pattern Anal. Mach. Intell. **31**(5), 855–868 (2009)
17. Sak, H., Andrew, Beaufays F.: Long short-term memory recurrent neural network architectures for large scale acoustic modeling (2014)
18. Li, X., Wu, X.: Constructing long short-term memory based deep recurrent neural networks for large vocabulary speech recognition (2014)

19. Hochreiter, S., Schmidhuber, J.: Long short-term memory. Neural Comput. **9**(8), 1735–1780 (1997)
20. Cho, K., van Merrienboer, B., Gulcehre, C., Bahdanau, D., Bougares, F., Schwenk, H., Bengio, Y.: Learning phrase representations using RNN encoder-decoder for statistical machine translation (2014)
21. Comodi, G., Renzi, M., Cioccolanti, L., Caresana, F., Pelagalli, L.: Hybrid system with micro gas turbine and PV (photovoltaic) plant: guidelines for sizing and management strategies. Energy. **89**, 226–235 (2015)
22. Serale, G., Fiorentini, M., Capozzoli, A., Bernardini, D., Bemporad, A.: Model predictive control (MPC) for enhancing building and HVAC system energy efficiency: problem formulation, applications and opportunities. Energies. **11**, 631 (2018)
23. Acosta, A., González, A., Zamarreño, J., Álvarez, V.: Energy savings and guaranteed thermal comfort in hotel rooms through nonlinear model predictive controllers. Energy Build. **129**, 59–68 (2016)
24. Riverón, I., Gómez, J.F., González, B., Méndez, J.A.: An intelligent strategy for hybrid energy system management. Renew. Energy Power Qual. **17**, 5 (2019)
25. Gers, F.A., Schmidhuber, J., Cummins, F.: Learning to forget: continual prediction with LSTM. Neural Comput. **12**(10), 2451–2471 (2000)
26. Million, E.: The hadamard product (2007)
27. Gers, F., Schmidhuber, J., Cummins, F.: Learning to forget: continual prediction with LSTM. In: 1999 Ninth International Conference on Artificial Neural Networks ICANN 99, IEEE, pp. 850–855 (1999)
28. Chung, J.; Gulcehre, C., Cho, K., Bengio, Y.: Empirical evaluation of gated recurrent neural networks on sequence modeling (2014)
29. Kruskal-Wallis, H.: Test using SPSS Statistics, Laerd Statistics
30. Lowry, R.: One way ANOVA - independent samples
31. Holm, S.: A simple sequentially rejective multiple test procedure. Scandinavian J. Stat. **6**(2), 65–70 (1979)

Special Session on Mathematical Techniques in Artificial Intelligence and Machine Learning (MaTe-AI&ML)

This special session is devoted to gather high-quality research papers and reviews focused on the use, study, and analysis of mathematical and statistical techniques in artificial intelligence and machine learning.

Specifically, this special session will cover different perspectives on these and related potential topics:

- Mathematical foundations of computer science.
- Mathematical foundations of artificial intelligence and machine learning.
- Mathematical foundations of data science and big data.
- Computational modeling and simulation in computer science, artificial intelligence, machine learning, and data science.
- Emerging methodologies, technologies, and applications in computer science, artificial intelligence, machine learning, and data science.
- Formal languages and automatas
- Mathematical foundations of cryptography and cybersecurity
- Quantum computing
- Mathematical modeling and simulation of complex systems and intelligent systems
- Soft computing including fuzzy sets
- Optimization, control, and modeling of processes and procedures.
- Fault detection and diagnosis.

Organizing Committee

Angel Martin del Rey, Universidad de Salamanca, Spain
Roberto Casado-Vara, University of Burgos, Spain

Explainable Artificial Intelligence on Smart Human Mobility: A Comparative Study Approach

Luís Rosa[1](✉), Fábio Silva[1,2], and Cesar Analide[1]

[1] Department of Informatics, ALGORITMI Center, University of Minho, Braga, Portugal
{id8123,analide}@di.uminho.pt, fas@estg.ipp.pt
[2] CIICESI, ESTG, Politécnico do Porto, Felgueiras, Portugal

Abstract. Explainable artificial intelligence has been used in several scientific fields to understand how and why a machine learning model makes its predictions. Its characteristics have allowed for greater transparency and outcomes in AI-powered decision-making. This building trust and confidence can be useful in human mobility research. This work provides a comparative study in terms of the explainability of artificial intelligence on smart human mobility in the context of a regression problem. Decision Tree, LIME, SHAP, and Seldon Alibi are explainable approaches to describe human mobility using a dataset generated from New York Services. Based on our results, all of these approaches present relevant indicators for our problem.

Keywords: Explainable artificial intelligence · Machine learning · Smart cities · Smart human mobility

1 Introduction

Machine Learning (ML) algorithms have been used in many application domains. These algorithms are being employed to complement humans' decisions in various tasks from diverse domains, such as finance, travel and hospitality, law enforcement, health care, news and entertainment, logistics, and manufacturing [12]. However, they still face acceptability issues. The major disadvantage of Deep Learning (DL) remains that its numerous parameters are challenging to interpret and explain. This especially holds true for opaque decision-making systems which are considered complex black box models.

The inability for humans to see inside black boxes can result in an increased need for interpretability, transparency, and explainability of AI products/outputs like predictions, decisions, actions, and recommendations. These elements are required to ensure the explanation of ML decisions or functionality. In [6], Explainable Artificial Intelligence (XAI) refers to techniques and methods of explaining so that AI solution results can be understood by humans.

J. M. Machado et al. (Eds.): DCAI 2022, LNNS 585, pp. 93–103, 2023.
https://doi.org/10.1007/978-3-031-23210-7_9

On human mobility context some XAI projects have been studied. For example, [7] focuses on the topic of automatic detection of Search and Rescue (SAR) missions, by developing and evaluating a methodology for classifying the trajectories of vessels that possibly participate in such missions. Luca et al. provides a taxonomy of mobility tasks where a discussion on the challenges related to each task and how deep learning may overcome the limitations of traditional models and a description of the most relevant solutions to the mobility tasks are described [9]. However, unfortunately, scarce contributions have been made to explore the human mobility field from a conceptual perspective.

This paper aims at filling this gap by focusing on human mobility, from a conceptual and practical point of view, and proposes approaches on how to evaluate XAI methods. This project brings together human mobility and explainable AI algorithms like LIME, Decision Tree, SHAP and Seldon Alibi in the same discussion. Therefore, a conceptual framework is built based on the foundation of the proposed goal methods for explainability. We also introduce models using explanators where these can be evaluated by employing notions and metrics.

The rest of the paper is organized as follows. In Sect. 2, it compares Classification and Regression projects and defines the concept of four interpretability techniques. XAI methods in human mobility are discussed in Sect. 3. Then, in Sect. 4, we discuss the output created by XAI algorithms considering the proposed data. Finally, this paper shows that interpretable AI can be effectively used for future human mobility surveys.

2 Explainable Artificial Intelligence and Human Mobility Research

Based on the literature, Explainable Artificial Intelligence (XAI) is applied in several studies that address classification and regression problems. Many of these works analyze prediction of human tracking, looking at a large population of free-will and autonomous decision-making individuals, or at any event that implies a restriction in mobility. In the following subsections, we indicate relevant researches on human mobility and the challenges and opportunities that they bring to XAI area.

2.1 Classification *Versus* Regression problems

Some studies have been conducted using classification algorithms. For example, [4] englobes 5G and 6G and it needs sophisticated AI to automate information delivery simultaneously for mass autonomy, human machine interfacing, and targeted healthcare. The survey analyses the results of three XAI methods, including LIME, SHAP, and CAM. Cao et al. [5] finds that LIME's explanation shows the most influenced image regions on the image classification problem's prediction. In its turn, [14] provides a review on interpretabilities suggested by different research works and categorizes them. The authors compare a set of xAI methods such as Layer-wise Relevance Propagation (LRP), LIME and others to ensure mainly that clinicians and practitioners can subsequently approach these methods with caution and insights into interpretability born with more considerations for medical practices.

On the other hand, regression problems have also been developed. In [11], researchers design hybrid models, combining the expressiveness of opaque models with the clear semantics of transparent models where linear regression is combined with neural networks. Then, [8] develops a ML algorithm using the XGBoost technique, and feature importance, partial dependence plot, and Shap Value is used to increase the model's explanatory potential. Adadi and Berrada [1] provides an entry point for interested researchers and practitioners to learn key aspects of the recent and rapidly growing body of research related to XAI. In short, we provide several illustrative classification and regression examples from various fields, motivating the need for a distinct treatment of both explanation problems respectively.

2.2 XAI Research Methods in Human Mobility

Through the lens of the literature, we review the existing approaches regarding the XAI topic and present trends surrounding it. These elements should be further analysed in human mobility context. In other words, an explainability project that aggregates a discussion human mobility, XAI and their methods should be considered to understand ML algorithms perform so that governments or local authorities understand and solve the diverse pedestrian problems in smart cities. For that reason, we now synthesize and enumerate a set of XAI methods useful for our regression problem:

- Decision Tree - This approach/technique is based on Classification and Regression Trees (CART) which deal with all kinds of variables and predict both numerical and categorical attributes. We use a Decision Tree (DT) API developed by [3].
- LIME - This framework also generates prediction explanations for any classifier or ML regressor. Its main advantage is the ability to explain and interpret the results of models using text, tabular and image data.
- SHAP - This approach is based on a game theory to explain the output of ML models. It provides a means to estimate and demonstrate how each feature's contribution influence the model.
- Seldon Alibi - An open source Python library aimed at ML model inspection and interpretation. Its focus is to provide high-quality implementations of black-box, white-box, local and global explanation methods for classification and regression models.

Based on previous frameworks we have built a transparent, trusted AI to instil trust among our work. They are one of the key requirements for implementing responsible AI in human mobility problems. In the following section we present a solid understanding of how to compute and interpret them in our case study.

3 Analysis of Human Mobility via XAI

In this section, we take a practical hands-on approach, using proposed XAI methods on Sect. 2.2. But first of all we outline considerations for analysing aggregated data from several open-source APIs made available by New York City authorities. Through these services we define the dataset for this work and when applied with these methods, we obtain important explanations about human mobility phenomena.

3.1 Data Collection

The data collection of this work is generated via a free public data published by New York agencies and other partners. However, much of this available data is scattered in several API services. In order to aggregate data from these open sources we develop a script in Python language. It uses operations Socrata Open Data API, developed and managed by the Department of Information Technology and Telecommunications (DoITT). Subsequently, we build the Python application taking into account the following sources:

- LinkNYC Kiosk Status - This application provides the most current listing of kiosks, their location, and the status of the Link's WiFi, tablet, and phone;
- 311 Service Requests - The 311 line provides its residents with a resource for assistance and general information outside of emergency situations;
- TLC Trip Record Data - The yellow and green taxi trip records are collected and provided to the NYC Taxi and Limousine Commission by technology providers authorized under the Taxicab & Livery Passenger Enhancement Programs.

We also add the *sentiment* attribute to the generated dataset from NYC APIs. It is calculated from a sentiment analysis process using Valence Aware Dictionary and Sentiment Reasoner (VADER). This "computationally" determines whether a piece of writing is positive (i.e., value 1), neutral (i.e., value 0), or negative (i.e., value -1) based on the *description* or *report* attribute. Additionally, weather information also enriches the original data, and for this purpose we use the OpenWeatherMap API. It allows collecting a vast amount of data associated with weather conditions such as clouds, feels like, humidity, pressure, temperature, temperature minimum, temperature maximum and speed [15].

Once the services are synced, data is stored on a PostgreSQL database. Then, we aggregate the rows of original table via SQL query, reducing memory consumption and processing time. Finally, with download via REST API, we have our dataset.

3.2 Exploring XAI Methods

As we demonstrated in Sect. 2.2, a set of xAI methods with a mutual case study/task (i.e., census prediction) is defined to analyze and provide meaningful insight on quantifying explainability, and recommend paths towards human mobility. Therefore, we leverage more detailed information about its implementation on our work.

Decision Tree API Inspired by the ML library, scikit-learn has an optimized version of the CART algorithm despite not yet supporting categorical variables [2]. Due to the explanatory potential of CART algorithm algorithm, Miguel Guimarães et al. implements a scratch of an Decision Tree (DT) [3]. The developed DT script is open-source where any programmer can download the source code and modify it, without necessarily depending on the API.

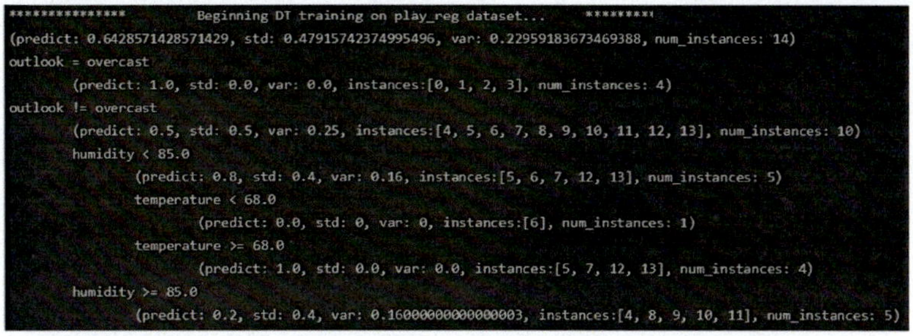

```
***************          Beginning DT training on play_reg dataset...     *********
(predict: 0.6428571428571429, std: 0.47915742374995496, var: 0.22959183673469388, num_instances: 14)
outlook = overcast
        (predict: 1.0, std: 0.0, var: 0.0, instances:[0, 1, 2, 3], num_instances: 4)
outlook != overcast
        (predict: 0.5, std: 0.5, var: 0.25, instances:[4, 5, 6, 7, 8, 9, 10, 11, 12, 13], num_instances: 10)
        humidity < 85.0
                (predict: 0.8, std: 0.4, var: 0.16, instances:[5, 6, 7, 12, 13], num_instances: 5)
                temperature < 68.0
                        (predict: 0.0, std: 0, var: 0, instances:[6], num_instances: 1)
                temperature >= 68.0
                        (predict: 1.0, std: 0.0, var: 0.0, instances:[5, 7, 12, 13], num_instances: 4)
        humidity >= 85.0
                (predict: 0.2, std: 0.4, var: 0.16000000000000003, instances:[4, 8, 9, 10, 11], num_instances: 5)
```

Fig. 1. The generated Decision Tree from the command line.

Either through the API or the execution of the DT script, the model training is not allowed if the settings given by the user are invalid. The user is also informed if there are possible improvements to be made in the settings. An example of running DT through the Command-Line Interface (CLI) is shown in Fig. 1.

LIME This XAI method manipulates the input data and creates a series of artificial data containing only a part of the original attributes [13]. It also provides local model interpretability. Basically, this explainable technique modifies a single data sample by tweaking the feature values and observes the resulting impact on the output. Moreover, its main characteristic is explaining to the dataset level which features are important. It fit the model using sample data points that are similar to the observation being explained. The explanations provided by LIME for each observation x is obtained as Eq. 1:

$$\Phi(x) = argmin_{g \in G} L(f, g, \pi_x) + \Omega(g) \tag{1}$$

where G is the class of potentially interpretable models such as linear models and decision trees, $g \in G$: an explanation considered as a model. $\pi_x(z)$ is proximity measure of an instance z from x. $\Omega(g)$ is a measure of complexity of the explanation $g \in G$.

The goal is to minimize the locality aware loss L without making any assumptions about f, since a key property of LIME is that it is model agnostic. L is the measure of how unfaithful g is in approximating f in the locality defined by $\pi(x)$.

SHAP Shap library is a tool proposed by Lundberg and Lee [10]. It adapts a concept coming from game theory and has many attractive properties. Additionally, we can "debug" our model and observe how it predicted an observation. In the most general form, the Shapley value is the average marginal contribution of a feature value across all possible coalitions. If there are N features, Shapley values will be computed from N different order combinations. From a computational perspective, has shown that the only additive method that satisfies the properties of local accuracy, missingness and consistency is obtained attributing to each variable $x'i$ an effect i defined by Eq. 2:

$$\phi_i(f, x) = \sum_{z' \subseteq x'} \frac{|z'|!(M - |z'| - 1)!}{M!} [f_x(z') - f_x(z' \backslash i)] \tag{2}$$

where f is the model, x are the available variables, and x are the selected variables. The quantity $f_x(z') - f_x(z' \backslash i)$ expresses, for each single prediction, the deviation of Shapley values from their mean: the contribution of the $i - th$ variable.

Seldon Alibi This XAI technique is an open source Python library aimed at Machine Learning (ML) model inspection and interpretation. It provides high-quality implementations of black-box and local explanation methods for regression models. Alibi currently ships with 8 different algorithms for model explanations including popular algorithms like anchors, counterfactuals, integrated gradients, Kernel SHAP, and Tree SHAP. For our work, we choose the Kernel SHAP.

When Alibi is invoked, training data is required and the fit method must be called. Then, an explain method is called to calculate an explanation on an instance or a set of instances. This returns an Explanation object containing dictionaries meta and data with the explanation metadata (e.g. hyperparameter settings, names) and the explanation data respectively. The structure of the Explanation object enables easy serialization in production systems for further processing (e.g. logging, visualization).

4 Results and Discussion

This guide is a practical instruction on how to use and interpret our models explainability. Our first explainable Machine Learning (ML) algorithm is Decision Tree (DT). It uses a linear model, it is also a relatively simple model, and explained by visualizing the tree represented in Fig. 2.

In this XAI technique, we see a sample forecast path from the root node to the leaf node generated via DT from multivariate model outcome. In root, the

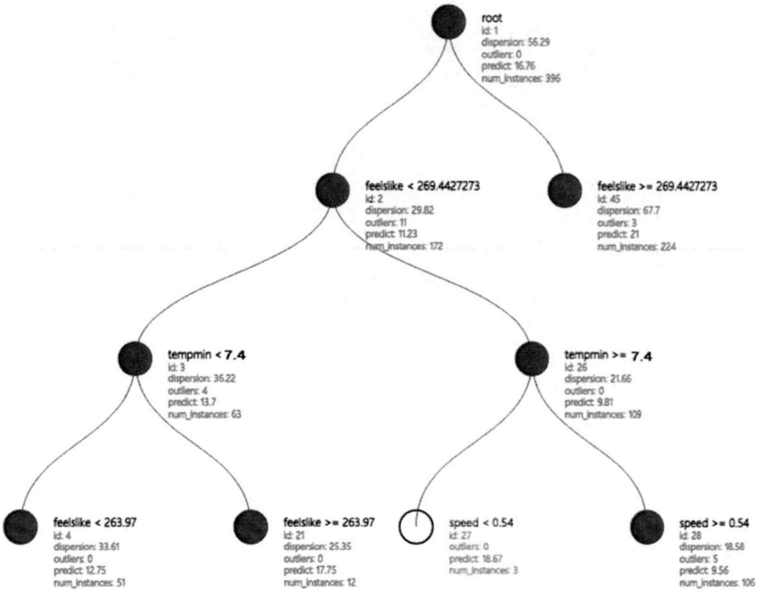

Fig. 2. The Decision Tree explanation.

DT predicts 16 individuals on NYC center. In the next forecast, with *feelslike* above or equal 269.442, DT predicts a number of 21, but argues the prediction based on the *feelslike* values below 269 with a slight error, 3 attributes (i.e., *feelslike*, *tempmin* and *speed*) are used to model this particular problem. In addition to this explanation, it also generates automatic counterfactual analysis. When the value of *tempmin* attribute is below 7.4 (°C), tree gives a tree level and predicts 14 people in New York City center. In case of three level, *feelslike* attribute predicts 13 people, otherwise 18 people. In its turn, when *tempmin* is above or equal 7.4 (°C) it also has an impact on census prediction, estimating 10 people. On child nodes, it should be noted that when *speed* value is below 0.54 the prediction of the number of people is 19. In *speed* value above or equal 0.54 the prediction is 10.

In its turn, the key functions in the LIME package are $LimeTabular$ $Explainer()$, which creates an explanation, and *explain_instance*(), which evaluates explanations. The *explain_instance* function requires three arguments: X_test, which specifies the test part of the first sequence (X), and predict function, which specifies the predict using the linear model. An additional, important argument is number features that indicates the maximum number of features present in explanation (K). Additionally, we specify that the number features of explanatory $num\_features$ variables is 3 (Fig. 3).

By applying the *as_pyplot_*$figure()$ to the object containing the explanation, we obtain a graphical presentation of the results. The output includes the colors blue and orange, depicting negative and positive associations, respectively. To interpret the above results, we can conclude that the relative census value

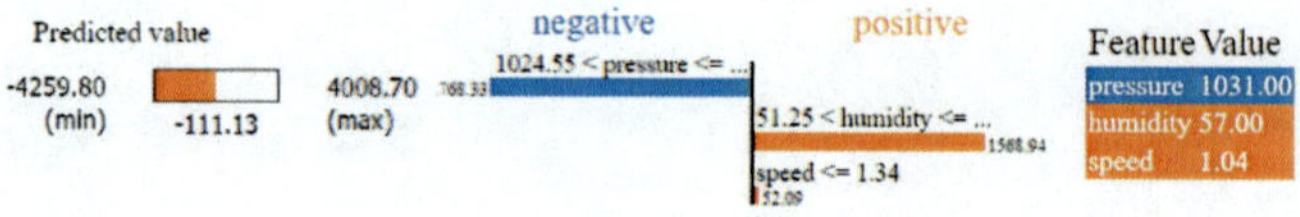

Fig. 3. The LIME explanation.

(depicted by a bar on the left) depicted by the given test vector (X) can be attributed to (1) the high value of *pressure* feature indicating the less number of people on New York City (NYC) center, (2) the high value of *humidity* feature indicating the high value of the number of people, and (3) the low value of *speed* of wind indicating the high value increase of people in the center.

First column *Predicted Value* computed value corresponds to the number given in the column *model_prediction* in the printed output. This value is approximately 111 individuals on NYC center. On the other hand, the *Feature Value* column or *model_intercept* column provides of the value of the intercept. It indicates which explanatory variables were given non-zero coefficients in the Linear Regression method. Additionally, it provides information about the values of the original explanatory variables for the observations for which the explanations are calculated.

SHAP framework has a SHAP Explainer that supports any and every ML algorithm. For instance, since we are handling a regression problem, which is bucketed under Linear model, we compute using a Linear Explainer. Thus, we use the *Explainer*() to build a new explainer for the passed model.

One the fundamental properties of Shapley values is all the input features will always sum up to the difference between expected model output and the current model output for the prediction being explained. Based on an observation, the easiest way to see this is through a waterfall plot that starts our background prior expectation for number of people on New York City center $E[f(X)]$. With features one at a time until we reach the current model output $f(x)$. In other words, this plot shows SHAP values for each of the features. Additionally, it tells how much each of the features have increased or decreased the predicted number of rings for this specific abalone.

Figure 4 shows expected value of the model output, and then each row shows how the positive (red) or negative (blue) contribution of each feature moves the value from the expected model output over the background dataset to the model output for this prediction.

Looking at the x-axis, we can see the base value is $E[f(x)] = 135.495$. This is the average predicted number of individuals on NYC center. The ending value is $f(x) = 407.4$. This is the predicted number of pedestrians in NYC center. The SHAP values are all the values in between. For example, the pressure the predicted number of people by 433.24 when compared to the average predicted census. Summarizing, each feature value shows how much each factor contributed to the model's prediction when compared to the mean prediction. Large positive/negative SHAP value indicate that the feature had a significant impact on the model's prediction.

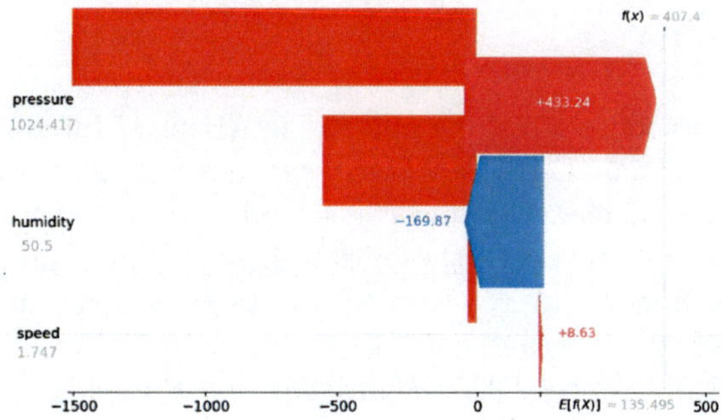

Fig. 4. The SHAP explanation.

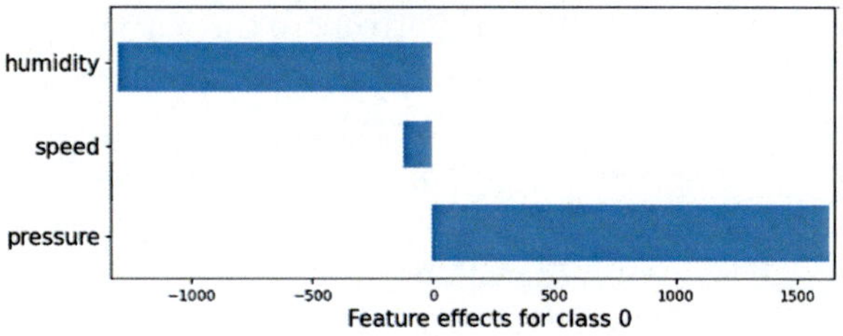

Fig. 5. The Seldon Alibi explanation.

An illustration of the Seldon Alibi explanation is shown in Fig. 5. This framework depicted a model which takes as an input features such as *Humidity*, *Speed* and *Pressure* and outputs a continuous value. We can see for example that the *Humidity* and *Speed* features contribute negatively to this prediction of census (i.e., number of individuals on NYC center) whereas the remainder of the feature have a positive contribution. To explain this particular data point, the *Pressure* feature seems to be the most important.

From the point of view of LIME and SHAP results, for the same model and the same data point, they provide different explanation. Although, they had the strongest impact in the prediction made, in LIME, the *Pressure*, *Humidity* and *Speed* features have a different priority than SHAP feature. But overall their explanations made sense. Unlike its predecessors, DT API simultaneously provides justification and attains with *Pressure* and *Speed* feature high accuracy, exactly opposite of *Humidity* feature. This allow for a new category of accuracy and interpretability. In the other hand, Seldon Alibi showed interesting results with Linear Regression model. Especially the class prototyping method is effective at accelerating counterfactual search, and in this case also improved the quality and plausibility of predictions. Only *Pressure* feature benefits a positive prediction result. However, regardless of the characteristics of previous XAI techniques, they reduce cost of wrong predictions and the impact of erroneous results, identifying the root cause leading to improving the underlying model.

5 Conclusions

In this work, we propose a set of Explainable Artificial Intelligence (XAI) methods like Decision Tree (DT), LIME, SHAP and Seldon Alibi. It is important to understand how each technique can be understood by humans, taking into account each of its features. In the initial phase, we presented a short intuition of each method, how to apply them to a dataset and compare the similarities between them and the pros and cons of each method for our problem. Comparing the performance of proposed XAI algorithm, the *Pressure*, *Humidity* and *Speed* attributes figure different results and protagonism, influencing the prediction value of each technique.

In future work, we plan to introduce more plots to show how each XAI technique contributes positively or negatively to the output of the model. Whilst, DT is visualized from impressive tree, LIME has another plot with a bar chart of local feature importance based on weights derived from Linear Regression. In case of SHAP we can also add a graph with the most relevant features in the highest positions and indicate how they affect the prediction. Lastly, we can introduce a chart that shows how the counterfactual methods change the features to flip the prediction in Seldon Alibi. Finally, another XAI method to explain our problem can be also analyzed such as InterpertML.

Acknowledgments. This work has been supported by FCT - Fundacao para a Ciencia e Tecnologia within the R&D Units Project Scope: UIDB/00319/2020. It has also been supported by national funds through FCT - Fundação para a Ciência e Tecnologia through project UIDB/04728/2020.

References

1. Adadi, A., Berrada, M.: Peeking inside the black-box: a survey on explainable artificial intelligence (XAI). IEEE Access **6**, 52138–52160 (2018). https://doi.org/10.1109/ACCESS.2018.2870052
2. den Bossche, J.V.: A new categorical encoder for handling categorical features in scikit-learn (2017). https://jorisvandenbossche.github.io/blog/2017/11/20/categorical-encoder/
3. Carneiro, D., Silva, F., Guimarães, M., Sousa, D., Novais, P.: Explainable intelligent environments. In: Advances in Intelligent Systems and Computing, vol. 1239 AISC, pp. 34–43. Springer, Cham (2021). https://doi.org/10.1007/978-3-030-58356-9_4
4. Guo, W.: Explainable artificial intelligence for 6G: improving trust between human and machine. IEEE Commun. Mag. **58**(6), 39–45 (2020). https://doi.org/10.1109/MCOM.001.2000050
5. Cao, H.Q., Nguyen, H.T.T., Nguyen, K.V.T., Nguyen, P.X.: A novel explainable artificial intelligence model in image classification problem (2021)
6. Kalyanathaya, K.P., Krishna Prasad, K.: A literature review and research agenda on explainable artificial intelligence (XAI). Int. J. Appl. Eng. Manag. Lett. 43–59 (2022). https://doi.org/10.47992/ijaeml.2581.7000.0119

7. Kapadais, K., Varlamis, I., Sardianos, C., Tserpes, K.: A framework for the detection of search and rescue patterns using shapelet classification. Future Internet **11**(9), 192 (2019). https://doi.org/10.3390/fi11090192
8. Lee, Y.: Applying explainable artificial intelligence to develop a model for predicting the supply and demand of teachers by region. J. Educ. e-Learn. Res. **8**(2), 198–205 (2021). https://doi.org/10.20448/journal.509.2021.82.198.205
9. Luca, M., Barlacchi, G., Lepri, B., Pappalardo, L.: A survey on deep learning for human mobility. ACM Comput. Surv. **55**(1), 1–44 (2023). https://doi.org/10.1145/3485125
10. Lundberg, S.M., Lee, S.I.: A unified approach to interpreting model predictions. In: Advances in Neural Information Processing Systems, vol. 2017-Dec, pp. 4766–4775. Neural Information Processing Systems Foundation (May 2017). https://doi.org/10.48550/arxiv.1705.07874
11. Munkhdalai, L., Munkhdalai, T., Ryu, K.H.: A locally adaptive interpretable regression (May 2020). https://doi.org/10.48550/arxiv.2005.03350
12. Rai, A.: Explainable AI: from black box to glass box. J. Acad. Market. Sci. **48**(1), 137–141 (2019). https://doi.org/10.1007/s11747-019-00710-5
13. Ribeiro, M.T., Singh, S., Guestrin, C.: "why should i trust you?" explaining the predictions of any classifier. In: NAACL-HLT 2016–2016 Conference of the North American Chapter of the Association for Computational Linguistics: Human Language Technologies, Proceedings of the Demonstrations Session, pp. 97–101. Association for Computational Linguistics (ACL) (Feb 2016). https://doi.org/10.18653/v1/n16-3020
14. Tjoa, E., Guan, C.: A survey on explainable artificial intelligence (XAI): toward medical XAI. IEEE Trans. Neural Netw. Learn. Syst. **32**(11), 4793–4813 (2021). https://doi.org/10.1109/TNNLS.2020.3027314
15. Weather Underground: OpenWeatherMap API (2018). https://www.wunderground.com/weather/api

Recurrent Neural Networks as Electrical Networks, a Formalization

Mariano Caruso[1,2](✉) and Cecilia Jarne[3,4]

[1] Fundación I+D del Software Libre–FIDESOL, Granada, Spain
[2] Universidad Internacional de La Rioja–UNIR, La Rioja, Spain
mcaruso@fidesol.org, mariano.caruso@unir.net
[3] Departamento de Ciencia y Tecnología, Universidad Nacional de Quilmes UNQ, Buenos Aires, Argentina
cecilia.jarne@unq.edu.ar
[4] CONICET, Buenos Aires, Argentina

Abstract. Since the 1980s, and particularly with the Hopfield model, recurrent neural networks or RNN became a topic of great interest. The first works of neural networks consisted of simple systems of a few neurons that were commonly simulated through analogue electronic circuits. The passage from the equations to the circuits was carried out directly without justification and subsequent formalisation. The present work shows a way to formally obtain the equivalence between an analogue circuit and a neural network and formalizes the connection between both systems. We also show which are the properties that these electrical networks must satisfy. We can have confidence that the representation in terms of circuits is mathematically equivalent to the equations that represent the network.

Keywords: RNN · Electrical networks · Formalization

1 Introduction

During the 1980s, and particularly since the Hopfield model [8], recurrent neural networks became a topic of great interest. In particular, latter with the works of Funahashi, Nakamura and Kurita [5,6,9], which made it possible to link neural networks with the description of dynamic systems, this research field began to establish as an area in itself. There are multiple papers on how neural networks are universal approximators, (i.e. they can approximate any continuous function). The proofs tell that neural networks can approximate any continuous function [12,13]. Any finite-time trajectory of a given $n-$dimensional dynamical system can be approximately realized by the internal state of the output units of a continuous-time recurrent neural network with $n-$output units, some hidden units, and an appropriate initial condition [6]. It was not until the last ten years that the current computing algorithms, the new hardware and the theory of neural networks allowed enormous developments in various areas related to natural

J. M. Machado et al. (Eds.): DCAI 2022, LNNS 585, pp. 105–114, 2023.
https://doi.org/10.1007/978-3-031-23210-7_10

language, dynamical systems, neurosciences and time series analysis using such networks [7,17].

However, with the hardware that was available at that time, the first works of neural networks consisted of simple systems of a few neurons that were commonly simulated through analogue electronic circuits. The passage from the equations to those circuits in most of the works of that time was carried out in a direct way, without much formalism. Mainly because the objective was to show how effectively these circuits could constitute rudimentary neural networks, mind that the analogue circuits allowed to simulate these systems. The behaviour of the systems was studied when the parameters of the network varied. Some current works in the area of electronics take up this idea and through circuit simulations, or the synthesis of analogue circuits, study the properties in systems with few neurons. One example are the transitions to chaotic systems [1,16].

On the other hand, dedicated circuits are used in another field called Neuromorphic engineering [10,11], which is also known as neuromorphic computing. This is the use of very large-scale integration (VLSI) systems containing electronic analogue circuits that mimic neuro-biological architectures related to the nervous system. A neuromorphic computer is called any device that uses physical artificial neurons to do computations. Recently the term neuromorphic has been used to describe analogue, digital, mixed-mode analogue/digital VLSI (and also software systems) that implement models of neural systems used to understand perception, motor control, or multisensory integration. The implementation of neuromorphic computing on the hardware level is realized through oxide-based memristors, spintronic memories, threshold switches, and transistors. Training software-based neuromorphic systems of spiking neural networks can be achieved using error backpropagation, e.g., using Python-based frameworks. The training algorithms of these systems are still complex, and it is difficult to control the network parameters.

Motivated by these developments and the gap in the literature about the formal aspects, the present work shows how to formally obtain the equivalence between an analogue circuit and a neural network, and formalizes the connection between both systems. We also show which are the properties that these electrical networks must satisfy. To the best of our knowledge, this is not explicitly found in the literature. The aim of the analysis is to explain the case of the linear network, meaning when the transfer function is the identity. This case is of interest since it has been used in various works, but also because it is often used to approximate nonlinear systems to first-order [14,15].

2 Notation and Dynamics

We have a set of $n-$*artificial neurons*, for each of these there is a dynamic quantity called *activity*, represented by a function $h_i : \mathcal{T} \longrightarrow \mathcal{H} \subseteq \mathbb{R}$ with $i \in I_n$ and a $\mathcal{T} \subseteq \mathbb{R}$ is the set of temporary labels. We will use the following compact notation to denote the discrete set n natural numbers $I_n = \{1, \cdots, n\} \subset \mathbb{N}$. We can arrange these n functions in a column vector $\boldsymbol{h} = (h_1, \cdots, h_i, \cdots, h_n)^{\mathrm{t}}$, where t denotes matrix transposition. The vector $\boldsymbol{h}$ represents the *state* of the activity of the network (formed by the n neurons) at that time t. On the other hand, there are a series of m input functions, $x_k : \mathcal{T} \longrightarrow \mathcal{X} \subseteq \mathbb{R}$ with $k \in I_m$, which can be arranged in a column vector $\boldsymbol{x} = (x_1, \cdots, x_k, \cdots, x_m)^{\mathrm{t}}$. For recurrent neural networks (RNN) the activity vector $\boldsymbol{h}$ satisfies:

$$\dot{\boldsymbol{h}}(t) = -\boldsymbol{\lambda}\,\boldsymbol{h}(t) + \boldsymbol{\sigma}(\boldsymbol{w}\,\boldsymbol{h}(t) + \widetilde{\boldsymbol{w}}\,\boldsymbol{x}(t)), \tag{1}$$

where $\dot{\boldsymbol{h}}$ represents the total derivative with respect to time in the usual sense. The diagonal matrix $\boldsymbol{\lambda}$ contains the inverses of the characteristic times (τ_k, $k \in I_n$), of the postsynaptic modes of each neuron, $\boldsymbol{\lambda}^{-1} = diag(\tau_1, \cdots, \tau_k, \cdots, \tau_n)$. In the case where the neural network is completely disconnected, both internally $\boldsymbol{w} = \boldsymbol{0}$ and externally $\widetilde{\boldsymbol{w}} = \boldsymbol{0}$, the activity of each neuron h_k of the network decays exponentially each with a characteristic time given by τ_k. The matrices $\boldsymbol{w}$ and $\widetilde{\boldsymbol{w}}$ are $n \times n$ and $n \times m$, respectively. The matrix elements $\boldsymbol{w}$, w_{ij}, contain the synaptic connections, similar to $\widetilde{\boldsymbol{w}}$, $\boldsymbol{\sigma} : \mathbb{R}^n \longrightarrow \mathbb{R}^n$ is a vector field of *activation*. Strictly speaking, these fields usually have their image in some compact set, since the activation of the neurons has a saturation behavior, the typical examples are *hyperbolic tangent*, *logistics*, etc., which satisfies $\boldsymbol{\sigma}(\boldsymbol{0}) = \boldsymbol{0}$ (because the activity cannot be revived instantly, that is, the result of activating a neuron with zero activity is null). Furthermore, each of its components is defined by applying a single nonlinear function $\sigma : \mathbb{R} \longrightarrow \mathbb{R}$. That is, given a vector $\boldsymbol{\xi} \in \mathbb{R}^n$, expressed in components as $\boldsymbol{\xi} = (\xi_1, \cdots, \xi_n)$ is has to $\boldsymbol{\sigma}(\boldsymbol{\xi}) = (\sigma(\xi_1), \cdots, \sigma(\xi_n))$, as usual. The activity state of the network is determined by (1), which is updated as a result of the interaction between them via $\boldsymbol{w}$, with the external signals $\boldsymbol{x}$ that intervene on the activity of neurons according to $\widetilde{\boldsymbol{w}}$, and together with some initial condition. We could write (1) compactly as:

$$\dot{\boldsymbol{h}}(t) = \boldsymbol{F}(\boldsymbol{h}(t), \boldsymbol{x}(t)). \tag{2}$$

There are two procedures that can be performed on this differential equation in order to say something about the behavior of the system models: discretization and linearization. The first procedure allows computing the model through an algorithm. The second will allow us to clearly find a linear electrical network that captures all the dynamics of the recurrent neural network. The order of these procedures does not alter the result, or in other words, the order in which they are applied is independent. Intuitively we can anticipate this result, since each procedure is introduced by a different member of (2), discretization is applied on the differential operator on the left, while linearization is done on the activation function on the right.

3 Linearization Process

As the nonlinear character of $\boldsymbol{F}$ is exclusively in $\boldsymbol{\sigma}$, linearizing is a procedure that has to do with the *activation field* and the neuronal activity itself, that is, the activation object, which we are interested in considering. Within this activation field $\boldsymbol{\sigma}$, let us now look at each function within $\sigma : \mathbb{R} \longrightarrow \mathbb{R}$, and suppose that $\sigma(\xi)$ is $k-$times differentiable at $\xi = 0$, by Taylor's theorem, there is a function *remainder* $R_k(\xi)$ that allows us to write as

$$\sigma(\xi) = \sigma(0) + (d_\xi \sigma(\xi)|_{\xi=0})\xi + \cdots + (d_\xi^{(k)} \sigma(\xi)|_{\xi=0})\xi^k + R_k(\xi)\xi^k. \tag{3}$$

Activation functions σ are usually chosen whose tangent line has a slope equal to 1 at $\xi = 0$, that is, the activation function near the origin resembles the identity function, let us remember that $\sigma(0) = 0$. Using all this and for ξ small enough we can approximate $\sigma(\xi) \simeq \xi$. Local linear approximations also represent an important building block for the analysis of the behaviour of more complex, nonlinear dynamical systems [4]. This procedure will be valid for **regimes** of low neuronal activity. We are not saying that the linear approximation is valid **only** in this regime, Only stating that in this regime, the intensity of neuronal activity is so weak that there is a formal procedure that justifies the linear approximation. In fact, this approximation was also used in the case of long times. We understand that the reason for such a thing can be justified from the differential equation and affirm that it is correct to assume a certain *neuronal not saturation* in the long term. By long-term, we mean that over time, both because the matrix $\boldsymbol{A}$ (which is diagonalizable) is such that all its eigenvalues have a real part less than 0 (this is what is called asymptotic stability) or it can always be considered that the neuronal activation function, which takes the weighted sum of the activity signals of each neuron, prepares the whole situation. In any case, the final destination of the neuronal activity is terminal, this is further ensured by the second principle of thermodynamics, in the sense that every entity at some point will have very poor neuronal activity. In this way, under linearization we will have $\boldsymbol{F}(\boldsymbol{h}(t), \boldsymbol{x}(t)) \simeq -\boldsymbol{\lambda}\boldsymbol{h}(t) + \boldsymbol{w}\,\boldsymbol{h}(t) + \widetilde{\boldsymbol{w}}\,\boldsymbol{x}(t)$ thus the differential Eq. (2) takes the form

$$\dot{\boldsymbol{h}}(t) = \boldsymbol{A}\,\boldsymbol{h}(t) + \widetilde{\boldsymbol{w}}\,\boldsymbol{x}(t), \tag{4}$$

where $\boldsymbol{A} = \boldsymbol{w} - \boldsymbol{\lambda}$. We are interested in studying the activity of each neuron $\{h_k\}_{k \in I_n}$ subject to two types of interactions due to the interconnection: *inter-neuronal* connection given the weights matrix $\boldsymbol{w}$ and a series of external excitation $\boldsymbol{x}$ affecting each neuron through $\widetilde{\boldsymbol{w}}$ matrix. Linear systems are made particularly attractive by the fact that their asymptotic behaviour can be understood in terms of the eigenvalues and eigenvectors of $\boldsymbol{A}$ [4].

4 Electrical Networks

We will identify the system described by (1), in particular its linearized form (4) with an electrical network. An electrical network (EN) is understood as the composition of an oriented graph, where each of its arcs has two associated functions of time: current and voltage, these functions are linked by Kirchhoff's laws and by the arc relations that arise from the graph that represents it and the interconnected electrical elements, e.g. resistors, inductors, capacitors, etc. [2]. Whenever the method of nodes, loops or pairs of nodes [2] is used to find the dynamics of the network, systems of linear differential equations of first order or integro-differential equations will be obtained. To be able to use these methods, it is necessary to know the electrical network entirety, i.e. the elements that compose it and their arrangement and interconnection. Since we do not have this information, the solution must be structured on a general circuit, i.e. without taking into account its graph, or the elements in each of its arcs. Therefore it is not possible to apply any of the network analysis methods until we find a particular network and its arc elements. The way to solve this apparent circular problem requires not the *analysis* of one circuit, but the *synthesis* of all circuits in a given preferential family. Such requirements can be fixed from other assumptions which we will see below.

We intend to distinguish or identify *generalized coordinates* in the electrical network such that the Eq. (4) are satisfied. Since that equation has n degrees of freedom, therefore we distinguish n local regions in the network from which voltages or currents can be measured. These regions are called ports: a pair of terminals that allow to exchange energy with the surroundings and have a given port-voltage and port-current. We conclude such an electrical network has n independent ports perfectly identified. The general structure that we propose is an electrical network composed of n dipoles listed as $\{\mathcal{N}_k\}_{k\in I_n}$, also called one−port networks, interconnected through a n−port interaction network $\mathcal{N}$ [2,3]. We see that each of one−port network introduces a port−voltage or port−current, that will be corresponds to a coordinate, h_k from $\boldsymbol{h}$ in (4), see Fig. 1. Since $\boldsymbol{w}$ is time−independent, then the dynamics of the power grid must be invariant under time translations, this implies that this part of the $\mathcal{N}$ network does not contain internal generators. In fact, by observing (4) we can conclude that such generators are well identified with the excitation signals $\boldsymbol{x}$. The initial conditions will be given on each dipole (or *one*−port network), and also in the external excitation $\boldsymbol{x}$, the energy initially contained by the interaction network $\mathcal{N}$ can be nonzero. In this way, the theory of multi−port networks can be perfectly used to synthesise $\mathcal{N}$ as an *active electrical network* [2]. In this scenario is possible to define a transfer matrix function of $\mathcal{N}$ written as the quotient between the Laplace transformation to certain output signals and the Laplace transformation of certain input signals. Depending on which signals are considered as inputs and outputs, there are four general representations: transmission, impedance, admittance, or hybrid. Note that the last case of hybrid representations of $\mathcal{N}$ is ruled out, otherwise, the input and output signals, voltages and currents, from different ports would be mixed, thus losing the possibility that each network $\{\mathcal{N}_k\}_{k\in I_n}$ represents, *per se*, one and only one of the coordinates $\{h_k\}_{k\in I_n}$ described by (4). In conclusion, we have interested in transmission, impedance or admittance representation of the n−port network $\mathcal{N}$. The energy is initially provided by the list of one−port networks $\{\mathcal{N}_k\}_{k\in I_n}$. The corresponding dynamics of an electric network is defined by the appropriate use of the Kirchhoff rules that take care of the topology of the network, schematically represented in Fig. 1. We have said that the generalized coordinates will be port-voltages or port-current of each of the n networks of the list $\{\mathcal{N}_k\}_{k\in I_n}$. Therefore, the n−port network $\mathcal{N}$ acts as an interaction in the sense that it physically interconnects the n dipoles networks. So the non-interaction case corresponds to disconnecting the $\mathcal{N}$ network. In a RNN this interaction-free responds to the fact that the weights matrix satisfies $\boldsymbol{w} = \boldsymbol{0}$ and the external excitation $\boldsymbol{x} = \boldsymbol{0}$ or its weights matrix $\widetilde{\boldsymbol{w}} = \boldsymbol{0}$, so that following (4) each neuron has an activity signal given by $h_k(t) = \alpha_k e^{-\lambda_k t}$, for a given $\lambda_k \in \mathbb{R}$. In other words neurons do not *see* each other. Mathematically, this is due because the evolution Eq. (4) take the form

$$\dot{\boldsymbol{h}}(t) = -\boldsymbol{\lambda}\boldsymbol{h}(t), \tag{5}$$

and given the matrix $\boldsymbol{\lambda}$ is diagonal the system of equations is uncoupled, the activity of the neurons is relegated to its initial condition and to a behavior that decays exponentially with a characteristic time matrix.

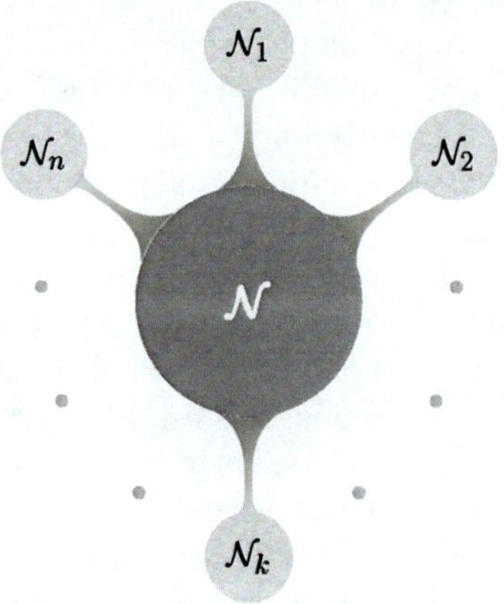

Fig. 1. There are n dipole networks, denoted by $\{\mathcal{N}_k\}_{k\in I_n}$ interconnected through an *interaction* network $\mathcal{N}$.

In order to compare directly with the result from the network synthesis method, let's apply the Laplace transform ($\mathcal{L}$) of the above linear differential Eq. (5) and taken the k−component of $\boldsymbol{h}(t)$ denoted by $h_k(t)$, then $H_k(s) = h_k(0)/(s + \lambda_k)$, where $H_k(s) = \mathcal{L}(h_k(t))_{(s)}$ and a nonnegative matrix $\boldsymbol{\lambda} = diag(\cdots, \lambda_k, \cdots)$. Note that each $H_k(s)$ is conceived as a characteristic function of a one-port $R_k \| C_k$ in parallel or one-port $R_k \boxplus L_k$ in series [2], these circuits are to be dual to each other. The identification ($\widehat{=}$) with the Eq. (4) is as follows: $\lambda_k^{-1} \widehat{=} R_k C_k$ and $h_k \widehat{=} v_k$ or $\lambda_k^{-1} \widehat{=} R_k/L_k$ and $h_k \widehat{=} i_k$. It should be noted that in each case the variable chosen is common to all of its elements: voltages and current for the parallel and series cases, respectively. Figure 2 summarize this situation. Note that the voltage source $v_k(0)$ and the current source $i_k(0)$ in Fig. 2 represent not only the initial condition but also mention that the initial energy is stored in the reactive elements. From an electromagnetic point of view, the initial potential difference in the capacitor C_k refers to the stored electrical energy given by $\frac{1}{2}C_k v_k^2(0)$. While the initial current in the inductor L_k refers to the stored magnetic energy given by $\frac{1}{2}L_k i_k^2(0)$. That is, both reactive elements are the initial source and thus provide the initial condition in each case.

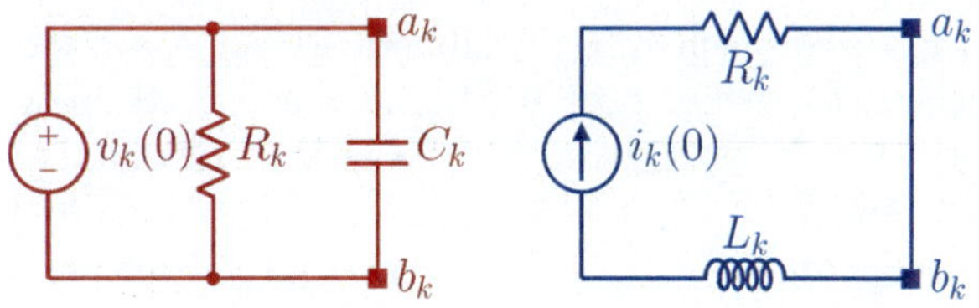

Fig. 2. Alternatives networks for each $\mathcal{N}_k$ in non-interaction case, i.e. a terminal dipole network of the list $\{\mathcal{N}_k\}_{k \in I_n}$ and its initial excitation, depending on which signal: voltages v_k (left and red) or currents i_k (right and blue) is chosen to describe the coordinate h_k of (4).

As we have said, the interaction of the components of neuronal activity vector $\boldsymbol{h}$ is due to an *internal* connection between neurons regulated by the internal weights matrix $\boldsymbol{w}$ and to an *external* connection regulated by the excitation $\boldsymbol{x}$ and the external weights matrix $\widetilde{\boldsymbol{w}}$. To establish the RNN correspondence with the electrical networks in this linear context, we need to connect the interaction network $\mathcal{N}$ in order to identify it with the matrices $\boldsymbol{w}$ and $\widetilde{\boldsymbol{w}}$ and the external excitation $\boldsymbol{x}$.

We claim that for a given a recurrent neural network regulated by (4), there are two electrical circuits, $\|$-*parallel* and $\boxplus$-*series* networks, that are dual to each other and reproduce the dynamics proposed by (5).

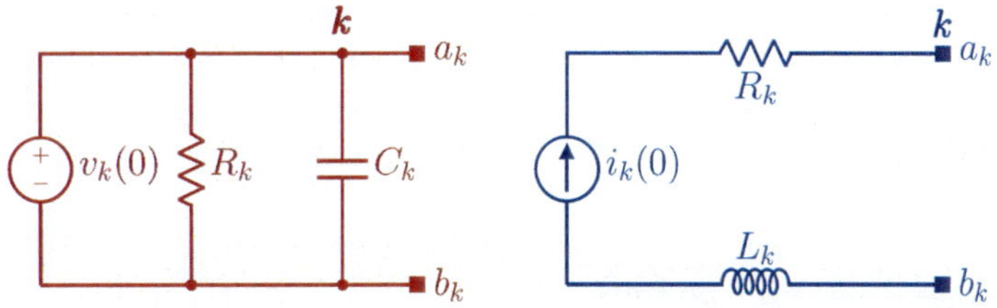

Fig. 3. Alternatives networks for $\mathcal{N}_k$, i.e. a terminal dipole network of the list $\{\mathcal{N}_k\}_{k \in I_n}$ and its initial excitation, depending on which signal: now are port–voltages v_k (left and red) or port–currents i_k (right and blue) i h_k of (4).

Note that the radical difference of each subnetwork $\mathcal{N}_k$ between the Figs. 2 and 3 is that the common signal v_k (i_k) for the parallel (serial) case is transmitted to the network $\mathcal{N}$; so each pair of terminals (a_k, b_k) are arranged to ensure this effect. To complete the electrical configuration of the complete network in Fig. 1, each pair of terminals (a_k, b_k) conforms a port that is connected to the k–port of $\mathcal{N}$. Depending on whether you choose to use a description in terms of voltages or currents, you will have to use an admittance or impedance representation for the associated $\mathcal{N}$ network.

To fix ideas we choose the *parallel network* and port–voltages of each subnetwork of the list $\{\mathcal{N}_k\}_{k\in I_n}$ as generalized coordinates, $\boldsymbol{v} = (v_1,\cdots,v_k,\cdots,v_n)$, at left on the Fig. 3. For each $k\in I_n$, a subnetwork $\mathcal{N}_k$ is a $R_k\|C_k$ tandem circuit, which is connected to the k–port of the network $\mathcal{N}$ as showed in Fig. 1. The k–node conform the k–port given by the pair of terminals (a_k, b_k), applying Kirchhoff's first law at this k–node: $i_{C_k} + i_{R_k} - i_k = 0$, using that $i_{R_k} = v_k/R_k$ and $i_{C_k} = C_k d_t v_k$, thus $C_k\dot{v}_k(t) + R_k^{-1}v_k(t) - i_k(t) = 0$, performing a Laplace transform ($\mathcal{L}$) then $C_k sV_k(s) - C_k v_k(0) + R_k^{-1}V_k(s) - I_k(s) = 0$. The last equation can be expressed in matrix form as

$$s\boldsymbol{V}(s) - \boldsymbol{v}(0) + \boldsymbol{\Lambda}\,\boldsymbol{V}(s) - \boldsymbol{C}^{-1}\boldsymbol{I}(s) = \boldsymbol{0}, \tag{6}$$

where $\boldsymbol{C} = diag(C_1,\cdots,C_n)$ and $\boldsymbol{R} = diag(R_1,\cdots,R_n)$, the matrix $\boldsymbol{\Lambda} = (\boldsymbol{RC})^{-1}$ contains the inverse of the characteristics times of each $R_k\|C_k$ subnetworks. In such a way, it all comes down to synthesizing the $\mathcal{N}$ network in the sense of the [2] circuit theory, in order to obtain a relation between the port-currents i_k and the port-voltages v_k. If the RNN does not have external excitation then the synthesis of the EN, $\mathcal{N}$, allows to express $\boldsymbol{I}(s) = \boldsymbol{Y}(s)\boldsymbol{V}(s)$, we have used the admittance representation of $\mathcal{N}$. Applying the inverse Laplace ($\mathcal{L}^{-1}$) transform to obtain the equation in the time domain

$$\dot{\boldsymbol{v}}(t) + \boldsymbol{\Lambda}\boldsymbol{v}(t) - \boldsymbol{C}^{-1}\mathcal{L}^{-1}[\mathbf{Y}(s)\boldsymbol{V}(s)](t) = \boldsymbol{0}. \tag{7}$$

The matrix elements of $\boldsymbol{Y}(s)$ are rational functions: quotients of polynomials in s. A necessary and sufficient condition for that the Eq. (7) has the form of (4) is $\boldsymbol{Y}(s) = \boldsymbol{\alpha}$, where the constant matrix $\boldsymbol{\alpha}$ of *conductances* can be synthesized using the general method exposed in [3]. There is non restrictions about the symmetry of the matrix $\boldsymbol{\alpha}$, in other words if we are interested in considering $\boldsymbol{\alpha}$ that is not necessarily symmetric, then the network $\mathcal{N}$ is said to be non-reciprocal. This implies that it can be synthesized using *gyrators* [3]. Comparing Eqs. (7) and (4) we obtain the identification $\boldsymbol{w} \hat{=} \boldsymbol{C}^{-1}\boldsymbol{\alpha} =: \boldsymbol{\Omega}$.

If we consider the complete description of RNN with the external excitation $\boldsymbol{x}$ then the electrical network $\mathcal{N}$ must be synthesized by $\boldsymbol{I}(s) = \boldsymbol{\alpha}\boldsymbol{V}(s) + \boldsymbol{\beta}\boldsymbol{U}(s)$, where $\boldsymbol{U}(s) = \mathcal{L}[\boldsymbol{u}(t)](s)$ and $\boldsymbol{u}(t) = (u_1,\cdots,u_m)$ are the voltages sources that act as the external excitation $\boldsymbol{x}$, in these case the weight matrix $\widetilde{\boldsymbol{w}} \hat{=} \boldsymbol{C}^{-1}\boldsymbol{\beta} =: \widetilde{\boldsymbol{\Omega}}$.

A similar procedure can be repeated in the impedance representation of the network $\mathcal{N}$ simply by interchanging the following quantities: voltages by currents, inductances by capacitances, conductances by resistances in order to obtain identical equations to (4) so that now the generalized coordinates are the port–currents $\boldsymbol{i}$.

The procedure we have described can be summarized in the following steps:

1. Propose the general topology of n sub-networks $\{\mathcal{N}_k\}_{k\in I_n}$ connected to a $\mathcal{N}$ network, hoping to be able to associate each artificial neuron in the RNN with a sub-network of $\{\mathcal{N}\}_{k\in I_n}$.

2. Identify the case of no interaction in both systems, in this case, RNN and EN.
3. Look for a dynamic quantity common to the, and representative of the sub-lattice $\mathcal{N}_k$ that corresponds to h_k. Note that it needs to be common and representative to capture that shared in $\mathcal{N}_k$ and to be able to identify the dynamics of each $\mathcal{N}_k$ with that of each h_k.
4. In the case considering interaction, this common and representative information of each $\mathcal{N}_k$ must be transferred to $\mathcal{N}$. This is achieved by transferring the potential difference (v_k) or the current flowing through the arc a_k and b_k (v_k) or the current between a_k and b_k (by previously opening such terminals) to the k−gate of $\mathcal{N}$.

The dynamics of the Electrical Network follows the differential equation

$$\dot{\boldsymbol{v}}(t) = -\boldsymbol{\Lambda}\boldsymbol{v}(t) + \boldsymbol{\Omega}\,\boldsymbol{v}(t) + \widetilde{\boldsymbol{\Omega}}\,\boldsymbol{u}(t) = \mathbf{0}, \tag{8}$$

we summarize the identification of the elements of an RRN (4) and the class of Electrical Networks (8) under study: $\boldsymbol{h} \hat{=} \boldsymbol{v}$, $\boldsymbol{\lambda} \hat{=} \boldsymbol{\Lambda}$, $\boldsymbol{\omega} \hat{=} \boldsymbol{\Omega}$, $\widetilde{\boldsymbol{\omega}} \hat{=} \widetilde{\boldsymbol{\Omega}}$ and $\boldsymbol{x} \hat{=} \boldsymbol{u}$.

For the nonlinear case, where the activation function σ plays am essential role, then we must consider the use of nonlinear amplifiers synthesis method with feedback in $\mathcal{N}$.

5 Discussion

It is well known that any finite-time trajectory of a given n−dimensional dynamical system can be approximately realized by the internal state of the output units of a continuous-time recurrent neural network with $n$output units. From this idea, and with the advance of the last ten years which includes current computing algorithms, the new hardware and the theory of neural networks, we have enormous developments in various areas related to natural language, dynamical systems, neuroscience and time series analysis.

While it may seem like an unnecessary step, being able to formalize and ground fundamental connections that are directly used from the very beginning allows us to learn more about systems in the process. It allows us to contextualize the types of circuits used and identify their characteristics and the parameters of the recurrent networks.

We have also carried out a review that allows us to present the current state of the art in the field of recurrent neural networks. We can perform simulations where we have an accurate representation of the phenomena associated with these systems. We have now the confidence that the representation is mathematically equivalent to the equations that represent the network.

6 Conclusions and Future Work

Since the idea of the current work was to formalize this equivalence implemented in the last 30 years, the objective was the development of such formalism in present paper. We have presented a procedure summarized in 4 steps to identify the elements of the electrical network with elements of the equation that represents the recurrent neural network. We think that a future work could address how to include specific conditions that neural networks must meet, such as Dale's law, or other constraints of biological origin and how they can affect the parameters of the circuits that emulate the networks.

Acknowledgements. Present work was supported by FIDESOL, CONICET and UNQ.

References

1. Ansari, M.S., Rahman, S.A.: DVCC-based non-linear feedback neural circuit for solving system of linear equations. Circ. Syst. Sig. Process. **30**(5), 1029–1045 (2011)
2. Balabanian, N., Bickart, T.A.: Linear Network Theory: Analysis, Properties. Design and Synthesis. Weber Systems (1982)
3. Carlin, H., et al.: Network Theory: An Introduction to Reciprocal and Nonreciprocal Circuits. Prentice-Hall Series in Electrical Engineering. Prentice-Hall (1964)
4. Duncker, L., Sahani, M.: Dynamics on the manifold: identifying computational dynamical activity from neural population recordings. Curr. Opin. Neurobiol. **70**, 163–170 (2021)
5. Funahashi, K.: On the approximate realization of continuous mappings by neural networks. Neural Networks **2**(3), 183–192 (1989)
6. Funahashi, K., Nakamura, Y.: Approximation of dynamical systems by continuous time recurrent neural networks. Neural Networks **6**(6), 801–806 (1993)
7. Gerstner, W., Sprekeler, H., Deco, G.: Theory and simulation in neuroscience. Science **338**(6103), 60–65 (2012)
8. Hopfield, J.J.: Neurons with graded response have collective computational properties like those of two-state neurons. Proc. Natl. Acad. Sci. **81**(10), 3088–3092 (1984)
9. Kurita, N., Funahashi, K.: On the Hopfield neural networks and mean field theory. Neural Networks **9**(9), 1531–1540 (1996)
10. Mead, C.: Neuromorphic electronic systems. Proc. IEEE **78**(10), 1629–1636 (1990)
11. Monroe, D.: Neuromorphic computing gets ready for the (really) big time. Commun. ACM **57**(6), 13–15 (2014)
12. Schäfer, A.M., et al.: Recurrent neural networks are universal approximators. In: Artificial Neural Networks—ICANN 2006, pp. 632–640. Springer, Berlin (2006)
13. Siegelmann, H.T., et al.: On the computational power of neural nets. In: Proceedings of the Fifth Annual Workshop on Computational Learning Theory, COLT'92, pp. 440–449. Association for Computing Machinery, New York, NY, USA (1992)
14. Sussillo, D.: Neural circuits as computational dynamical systems. Curr. Opin. Neurobiol. **25**, 156–163 (2014)
15. Sussillo, D., Barak, O.: Opening the black box: low-dimensional dynamics in high-dimensional recurrent neural networks. Neural Comput. **25**(3), 626–649 (2013)
16. Tabekoueng Njitacke, Z., Kengne, J., Fotsin, H.B.: Coexistence of multiple stable states and bursting oscillations in a 4d Hopfield neural network. Circ. Syst. Sig. Process. **39**(7), 3424–3444 (2020)
17. Trischler, A.P., D'Eleuterio, G.M.: Synthesis of recurrent neural networks for dynamical system simulation. Neural Networks **80**, 67–78 (2016)

Special Session on New Perspectives and Solutions in Cultural Heritage (TECTONIC)

The special session entitled "New perspectives and solutions in Cultural Heritage" is a forum that will share ideas, projects, researches results, models, experiences, applications, etc., and focus on Preventive Conservation of marine and aerial cultural heritage. New technological proposals, many of them based on artificial intelligence, are currently being applied as measures and processes for the conservation of cultural heritage. This special session has its main objectives:

• To maximize the value of research outcomes by promoting knowledge exchange, interactions, partnerships, and inclusive engagement between cultural heritage researchers, individuals, and organizations outside the immediate research community.

• To encourage the implementation and transmission of research outcomes and to communicate them and the knowledge acquired among researchers and stakeholder sectors.

The session was held in L'Aquila (Italy) as part of the 19th International Conference on Distributed Computing and Artificial Intelligence from July 13 to 15, 2022.

Organizing Committee

Mauro Francesco La Russa, University of Calabria, Italy
Michela Ricca, University of Calabria, Italy
Natalia Rovella, University of Bologna, Italy

Computer Vision: A Review on 3D Object Recognition

Yeray Mezquita[1(✉)], Alfonso González-Briones[1], Patricia Wolf[2], and Javier Prieto[1]

[1] BISITE Research Group, University of Salamanca, Edificio Multiusos I+D+i, Calle Espejo 2, 37007 Salamanca, Spain
yeraymm@usal.es, alfonsogb@usal.es, javierp@usal.es

[2] Department of Marketing & Management, University of Southern Denmark, Campusvej 55, 5230 Odense M, Denmark
pawo@sam.sdu.dk

Abstract. Three-dimensional (3D) object recognition is one of the fundamental tasks in computer vision applications, paving the way for all other image understanding operations. Although the potential of 3D object recognition is enormous, the information required for the spatial processing of the information means that the practical applications that end up being developed are very limited by the computational cost of the algorithms and frameworks used. This manuscript seeks to collect information on the most current review works in the literature. Thanks to this, researchers and developers who start working in the field of 3D object identification can find a compilation of the most important points to understand the current context in this field.

Keywords: Review · 3D object recognition · Computer vision

1 Introduction

Three-dimensional (3D) object recognition is one of the fundamental tasks in computer vision applications, paving the way for all other image understanding operations [9,13,19,25,26,36,41]. The demand for 3D object recognition is increasing due to the widespread of its applications in areas such as artificial intelligence robots, automated driving, medical image analysis, and virtual/augmented reality among other areas [3,18,20,21,23,24,32,35].

Although the potential of 3D object recognition is enormous, the information required for the spatial processing of the information means that the practical applications that end up being developed are very limited by the computational cost of the algorithms and frameworks used [40]. This is why over time more and more ways of working with three-dimensional image data have been developed to optimize the applications that require their use [29]. Besides, the need for better data bases that help in the creation of those frameworks and algorithms

J. M. Machado et al. (Eds.): DCAI 2022, LNNS 585, pp. 117–125, 2023.
https://doi.org/10.1007/978-3-031-23210-7_11

has driven the apparition of works like the one presented in [32], where the databases created are the main focus.

This manuscript seeks to collect information on the most current review works in the literature. Thanks to this, researchers and developers who start working in the field of 3D object identification can find a compilation of the most important points to understand the current context in this field: the algorithms and databases used to develop and evaluate them. Since this work is part of the "Technological Consortium TO develop sustainability of underwater Cultural heritage" project (TECTONIC), which makes use of 3D images of the seabed, we will also talk about related studies more specifically in this area.

The paper continues with an explanation of the 3D computer vision in Sect. 2. This is followed by the study model in the Sect. 3 followed by a discussion of the work found in the Sect. 4. Finally, the conclusions are detailed in the Sect. 5.

2 Background on 3D Object Identification

In this section it is going to be described a general background of the 3D computer vision field. 3D computer vision can be broadly defined as technologies that enable three-dimensional measurement or inspection of objects or surfaces in 3D. One of the advantages of 3D computer vision over 2D computer vision is that, in the case of occlusion, the views from different viewpoints can complement each other's detail features of the object and achieve excellent recognition performance. The data-sets used in 3D computer vision are created in different ways:

- Laser profiling: Laser profiling is one of the most popular 3D imaging techniques. The object being measured is moved through a laser beam as a camera positioned at a known angle records the changing profile of the laser as the object moves through it. This configuration is particularly popular on factory production floors or packaging lines as it relies on the movement of the object relative to the laser, meaning it is well suited to products on conveyor belts.
- Stereo imaging: Another popular 3D imaging technique is stereo imaging, in which two cameras are used to record 2D images of an object that can then be triangulated and converted into a 3D image. Like laser profiling, these techniques also allow for object movement during measurement and registration. The use of a random static illumination pattern can also give arbitrary texture to flat surfaces and objects that do not have natural edges, which many stereo reconstruction algorithms require.
- Fringe projection: In fringe projection, a fringe pattern is projected over the entire surface to be measured. The image is recorded by a camera positioned perpendicular to the object being measured. The point cloud created is capable of giving a height resolution up to two orders of magnitude higher than a laser profiling method is capable of providing. Fringe projection is also more scalable with a measurement area ranging from one millimeter to over one meter.

- Time of Flight: The time of flight method measures the time it takes for a pulse of light to reach the object being measured and then return. The time required to measure each point in the image will vary depending on the size and depth of the object and therefore each point will provide this information as they are measured.

There are different computer vision tasks, which can be classified as follows:

- Image classification. In this task, the image is assigned a label.
- Object localization. Once an object is located, a rectangle is drawn around it.
- Object detection. In this task the two previous tasks are combined, an object is located and a label is added to it.
- Instance segmentation. In this task, it is differentiated individuals of the same category.
- Semantic segmentation. Distinguishes high-level categories of significance, usually objects.
- Object recognition. Is a term that is used to referring all of the previous tasks together [30].

Model based methods can be divided in: i) voxel-based methods, where the objects can be represented as a 3D mesh, or ii) point-set-based methods, where a set of unordered points is used for prediction tasks [7,16,39,40].

3 Research Model

In this section, we will explain the development of the study model carried out in this work. To begin with, we want to answer the question: What has been published so far in the field of object identification in 3D images? Therefore, we have made use of the following macros: ("3D") AND ("object recognition") AND ("review").

Using the macros defined to search for academic articles in the ScienDirect database, we found 4764 results. Of which 3 have been used for this work:

- Review of multi-view 3D object recognition methods based on deep learning [29]. This manuscript presents an updated review and classification on deep learning methods for multi-view 3D object recognition. It also test these methods on mainstream datasets to provide some results and insights on the methods studied.
- Object recognition datasets and challenges: A review [32]. In this paper it is provided a detailed analysis of 160 datasets that has been widely used in the object recognition areas. It is also presented an overview of the object recognition benchmarks and the metrics adopted for evaluation purposes in the computer vision community.
- Review on deep learning techniques for marine object recognition: Architectures and algorithms [37]. The survey presented in this work is mainly focused in the marine object recognition methods based on deep learning for both,

surface and underwater targets. It is described typical deep network frameworks in three parts: image preprocessing, feature extraction, and recognition and model optimization.

4 Discussion

In this paper, we have surveyed some of the reviews in the literature on object recognition on 3D images. This work is necessary to find methods that have worked in this field and that can be used for the TECTONIC project, which will need to make use of these methods to label and identify objects in 3D images of the seafloor.

Based on what's said in [29], the use of deep learning techniques in multi-view 3D object recognition has become one of the most researched topics. It is because the deep learning techniques researched [2,6,17,22] can directly use the pretrained and successful advanced classification network as the backbone network, while thanks to the views obtained from multiple viewpoints it is possible to complement each one of the features of any object. There are still some challenges existing in this research topic, that's why many methods are being proposed in the literature to tackle them [8,15,28].

There are some works that propose the use of a mature classification network, such as VGG-M [4], VGG19 [33], GoogLeNet [34], AlexNet [14], ResNet-18/50 [10], and DenseNet [12]. The classification network is pretrained by the large-scale 2D datasets, to extract features at the view level, used then as the backbone network [38]. To provide accurate object classification, view-level features extracted need the use of feature fusion techniques. In the original work, the method used was a simple max-pooling of the features, but it ignored the relationship between the features. Other works of the literature proposed methods such as Recurrent Neural Networks (RNN) [1,27], Long Short-Term Memory (LSTM) [11], dynamic routing [31], Graph Convolutional Neural Network (GCN) [38]. In real-world applications, active camera viewpoints selection methods do the task of object recognition by optimizing the number of view inputs, and solving the occlusion problem, while reducing the cost of mobile robot [5].

From the study done in [29] it can be concluded that the existing 3D object recognition methods, see Table 1, have demonstrated their advantages although some aspects still need to be improved. E.g. passive selection methods are computationally intensive, limiting their performance in multi-view 3D object recognition. Another example is that the active views selection method requires all views to be inputted, from which arise practical problems such as occlusion. Because of these problems, advances in computer vision will require the automatic selection of the best viewing angle of the object [41]. A view to be selected as the best one should be the one with the most abundant image information and the highest distinguishability. This requirement will help the network to achieve the best recognition performance with as few views as possible while reducing the cost of the mobile robot and getting rid of the occlusion problem.

Table 1. Table summary of the techniques listed.

Type of method	Techniques
Convolutional neural networks	VGG-M [4] VGG19 [33] GoogLeNet [34] AlexNet [14] ResNet-18/50 [10] DenseNet [12]
Recurrent neural networks	Deep recurrent attention model (DRAM) [1] Visual attention model [27] Long short-term memory (LSTM) [11]
Capsules	Dynamic routing [31]
Graphs in CNN	View-graph convolutional neural network (GCN) [38]
Multi-view deep neural network	Veram [5]

In the manuscript [32] it's been shown the importance of the size and quality of datasets as the use of deep-learning techniques, which heavily rely on training data, spreads on the object recognition area. Also, the datasets are needed to provide a fair benchmarking mean for competitions, proving to be instrumental to the advancements in the field, by providing quantifiable benchmarks for the developed models. Also, it is being found the important of developing new and more challenging datasets, as the algorithms mature and the existing datasets become saturated. As a conclusion to the review, the authors state that researchers need to find the appropriate training and testing mediums for their desired applications.

In [37] authors exclusively focus on the deep-learning-based marine object recognition. According to this manuscript, marine object recognition propose various subproblems, such as the resolution (low, moderate and high) problem of images, samples starvation in image and video data, complex marine environmental factors, different degrees of model architectures, and optimizations based on the supervised and unsupervised learning models. Based on the literature review provided, and to provide guidelines for the researchers, they list the issues and challenges found as: i) the need to improve the public marine datasets, ii) necessity of pre-training, iii) the need of a unified framework, iiii) the fusion of multi-source features, v) fusion of multi-deep model, vi) sub-class recognition, and vii) the general model structure. Furthermore, it is proven that deep-learning methods, regardless of supervised or unsupervised, can be used in object recognition underwater and the surface.

5 Conclusion

In this article, we have conducted a study on the current context of computer vision for object recognition in three-dimensional environments. From the study, we found that deep learning techniques are the most widely used and investigated so far. They have proven to be effective and practical. Furthermore, we can conclude that the study, design, and development of new databases help enormously in this area since deep learning techniques are highly dependent on the data used and the current databases are starting to be saturated. Finally, we have studied the context in which computer vision is found in the maritime field. It can be said that object detection, both underwater and on the surface, is following the same development steps as in other fields, with deep learning techniques being the most widespread at present, which makes the development of new public databases for this field even more necessary.

The present study is far from perfect, it could be improved, not only by analyzing more works, but also to point out the advantages and disadvantages of each of the works in the literature. Besides, a comparison between the techniques, in different databases, applied to the maritime field could be a good way to differentiate the work while improving its scientific contribution.

Acknowledgements. The research of Yeray Mezquita is supported by the predoctoral fellowship from the University of Salamanca and co-funded by Banco Santander. This research was also partially supported by the project "Technological Consortium TO develop sustainability of underwater Cultural heritage (TECTONIC)", financed by the European Union (Horizon 2020 research and innovation programme under the Marie Skłodowska-Curie grant agreement No 873132). Authors declare no conflicts of interest.

References

1. Ba, J., Mnih, V., Kavukcuoglu, K.: Multiple object recognition with visual attention. arXiv preprint arXiv:1412.7755 (2014)
2. Casado-Vara, R., Novais, P., Gil, A.B., Prieto, J., Corchado, J.M.: Distributed continuous-time fault estimation control for multiple devices in IoT networks. IEEE Access **7**, 11972–11984 (2019)
3. Castellanos-Garzón, J.A., Mezquita Martín, Y., Jaimes, S.J.L., López, S.M.: A data mining approach applied to wireless sensor networks in greenhouses. In: International Symposium on Distributed Computing and Artificial Intelligence, pp. 431–436. Springer, Berlin (2018)
4. Chatfield, K., Simonyan, K., Vedaldi, A., Zisserman, A.: Return of the devil in the details: delving deep into convolutional nets. arXiv preprint arXiv:1405.3531 (2014)
5. Chen, S., Zheng, L., Zhang, Y., Sun, Z., Xu, K.: Veram: view-enhanced recurrent attention model for 3d shape classification. IEEE Trans. Vis. Comput. Graph. **25**(12), 3244–3257 (2018)
6. Fatima, N.: Enhancing performance of a deep neural network by comparing optimizers experimentally. ADCAIJ Adv. Distrib. Comput. Artif. Intell. J. **9**(2), 79–90 (2020). ISSN: 2255-2863

7. Feng, Y., Xiao, J., Zhuang, Y., Yang, X., Zhang, J.J., Song, R.: Exploiting temporal stability and low-rank structure for motion capture data refinement. Inf. Sci. **277**, 777–793 (2014)
8. González-Briones, A., Castellanos-Garzón, J.A., Mezquita Martín, Y., Prieto, J., Corchado, J.M.: A framework for knowledge discovery from wireless sensor networks in rural environments: a crop irrigation systems case study. Wirel. Commun. Mobile Comput. **2018** (2018)
9. Gupta, S., Meena, J., Gupta, O., et al.: Neural network based epileptic EEG detection and classification (2020)
10. He, K., Zhang, X., Ren, S., Sun, J.: Deep residual learning for image recognition. In: Proceedings of the IEEE Conference on Computer Vision and Pattern Recognition, pp. 770–778 (2016)
11. Hochreiter, S., Schmidhuber, J.: Long short-term memory. Neural Comput. **9**(8), 1735–1780 (1997)
12. Huang, G., Liu, Z., Van Der Maaten, L., Weinberger, K.Q.: Densely connected convolutional networks. In: Proceedings of the IEEE Conference on Computer Vision and Pattern Recognition, pp. 4700–4708 (2017)
13. Hussain, A., Hussain, T., Ali, I., Khan, M.R., et al.: Impact of sparse and dense deployment of nodes under different propagation models in manets (2020)
14. Krizhevsky, A., Sutskever, I., Hinton, G.E.: Imagenet classification with deep convolutional neural networks. In: Advances in Neural Information Processing Systems, vol. 25 (2012)
15. Li, T., Corchado, J.M., Sun, S.: Partial consensus and conservative fusion of gaussian mixtures for distributed PHD fusion. IEEE Trans. Aerosp. Electron. Syst. **55**(5), 2150–2163 (2018)
16. Liu, A.A., Zhou, H., Nie, W., Liu, Z., Liu, W., Xie, H., Mao, Z., Li, X., Song, D.: Hierarchical multi-view context modelling for 3d object classification and retrieval. Inf. Sci. **547**, 984–995 (2021)
17. López, A.B.: Deep learning in biometrics: a survey. ADCAIJ Adv. Distrib. Comput. Artif. Intell. J. **8**(4), 19–32 (2019)
18. Martín, Y.M., Parra, J., Pérez, E., Prieto, J., Corchado, J.M.: Blockchain-based systems in land registry, a survey of their use and economic implications. CISIS **2020**, 13–22 (2020)
19. Mezquita, Y., Alonso, R.S., Casado-Vara, R., Prieto, J., Corchado, J.M.: A review of k-nn algorithm based on classical and quantum machine learning. In: International Symposium on Distributed Computing and Artificial Intelligence, pp. 189–198. Springer, Berlin (2020)
20. Mezquita, Y., Casado-Vara, R., González Briones, A., Prieto, J., Corchado, J.M.: Blockchain-based architecture for the control of logistics activities: pharmaceutical utilities case study. Logic J. IGPL **29**(6), 974–985 (2021)
21. Mezquita, Y., Gil-González, A.B., Prieto, J., Corchado, J.M.: Cryptocurrencies and price prediction: a survey. In: International Congress on Blockchain and Applications, pp. 339–346. Springer, Berlin (2021)
22. Mezquita, Y., Gil-González, A.B., Martín del Rey, A., Prieto, J., Corchado, J.M.: Towards a blockchain-based peer-to-peer energy marketplace. Energies **15**(9), 3046 (2022)
23. Mezquita, Y., González-Briones, A., Casado-Vara, R., Chamoso, P., Prieto, J., Corchado, J.M.: Blockchain-based architecture: a mas proposal for efficient agri-food supply chains. In: International Symposium on Ambient Intelligence, pp. 89–96. Springer, Berlin (2019)

24. Mezquita, Y., González-Briones, A., Casado-Vara, R., Wolf, P., de la Prieta, F., Gil-González, A.B.: Review of privacy preservation with blockchain technology in the context of smart cities. In: Sustainable Smart Cities and Territories International Conference, pp. 68–77. Springer, Berlin (2021)
25. Mezquita, Y., Parra, J., Perez, E., Prieto, J., Corchado, J.M.: Blockchain-based systems in land registry, a survey of their use and economic implications. In: Computational Intelligence in Security for Information Systems Conference, pp. 13–22. Springer, Berlin (2019)
26. Mezquita, Y., Valdeolmillos, D., González-Briones, A., Prieto, J., Corchado, J.M.: Legal aspects and emerging risks in the use of smart contracts based on blockchain. In: International Conference on Knowledge Management in Organizations, pp. 525–535. Springer, Berlin (2019)
27. Mnih, V., Heess, N., Graves, A., et al.: Recurrent models of visual attention. In: Advances in Neural Information Processing Systems, vol. 27 (2014)
28. Pimpalkar, A.P., Raj, R.J.R.: Influence of pre-processing strategies on the performance of ml classifiers exploiting TF-IDF and bow features. ADCAIJ Adv. Distrib. Comput. Artif. Intell. J. **9**(2), 49 (2020)
29. Qi, S., Ning, X., Yang, G., Zhang, L., Long, P., Cai, W., Li, W.: Review of multi-view 3d object recognition methods based on deep learning. Displays **69**, 102053 (2021)
30. Russakovsky, O., Deng, J., Su, H., Krause, J., Satheesh, S., Ma, S., Huang, Z., Karpathy, A., Khosla, A., Bernstein, M., et al.: Imagenet large scale visual recognition challenge. Int. J. Comput. Vis. **115**(3), 211–252 (2015)
31. Sabour, S., Frosst, N., Hinton, G.E.: Dynamic routing between capsules. In: Advances in Neural Information Processing Systems, vol. 30 (2017)
32. Salari, A., Djavadifar, A., Liu, X.R., Najjaran, H.: Object recognition datasets and challenges: a review. Neurocomputing (2022)
33. Simonyan, K., Zisserman, A.: Very deep convolutional networks for large-scale image recognition. arXiv preprint arXiv:1409.1556 (2014)
34. Szegedy, C., Liu, W., Jia, Y., Sermanet, P., Reed, S., Anguelov, D., Erhan, D., Vanhoucke, V., Rabinovich, A.: Going deeper with convolutions. In: Proceedings of the IEEE Conference on Computer Vision and Pattern Recognition, pp. 1–9 (2015)
35. Valdeolmillos, D., Mezquita, Y., González-Briones, A., Prieto, J., Corchado, J.M.: Blockchain technology: a review of the current challenges of cryptocurrency. In: International Congress on Blockchain and Applications, pp. 153–160. Springer, Berlin (2019)
36. Vergara, D., Extremera, J., Rubio, M.P., Dávila, L.P.: The proliferation of virtual laboratories in educational fields. ADCAIJ Adv. Distrib. Comput. Artif. Intell. J. **9**(1), 85 (2020)
37. Wang, N., Wang, Y., Er, M.J.: Review on deep learning techniques for marine object recognition: architectures and algorithms. Control Eng. Pract. **118**, 104458 (2022)
38. Wei, X., Yu, R., Sun, J.: View-GCN: view-based graph convolutional network for 3d shape analysis. In: Proceedings of the IEEE/CVF Conference on Computer Vision and Pattern Recognition, pp. 1850–1859 (2020)
39. Yang, Z., Wang, L.: Learning relationships for multi-view 3d object recognition. In: Proceedings of the IEEE/CVF International Conference on Computer Vision, pp. 7505–7514 (2019)

40. Zhou, H.Y., Liu, A.A., Nie, W.Z., Nie, J.: Multi-view saliency guided deep neural network for 3-d object retrieval and classification. IEEE Trans. Multimedia **22**(6), 1496–1506 (2019)
41. Zubair, S., Al Sabri, M.A.: Hybrid measurement of the similarity value based on a genetic algorithm to improve prediction in a collaborative filtering recommendation system. ADCAIJ Adv. Distrib. Comput. Artif. Intell. J. **10**(2), 165–182 (2021)

An IoUT-Based Platform for Managing Underwater Cultural Heritage

Marta Plaza-Hernández(✉), Mahmoud Abbasi, and Yeray Mezquita

BISITE Research Group, University of Salamanca, Edificio Multiusos I+D+i, Calle Espejo 2, 37007 Salamanca, Spain
{martaplaza,yeraymm}@usal.es, Mahmoud.abbasi@ieee.org

Abstract. Conservation of Underwater Cultural Heritage is crucial to preserve society's history. This work proposes a platform based on Internet of Underwater Things technologies and the Edge Computing paradigm. It will incorporate Artificial Intelligence techniques that support the monitoring and management of Underwater Cultural Heritage. These algorithms and models, capable of generating knowledge, will work as a supporting tool for decision-making. The platform will integrate information stored in databases with data acquired in real-time, working independently and in collaboration with other platforms and systems.

Keywords: Underwater cultural heritage · Internet of underwater things · Artificial intelligence · Edge computing

1 Underwater Cultural Heritage: Threats and Challenges

As defined by the United Nations Educational, Scientific and Cultural Organization (UNESCO) in its 2001 Convention on the Protection of the Underwater Cultural Heritage (Art. 1.1(a)), "Underwater Cultural Heritage (UCH) means all traces of human existence having a cultural, historical or archaeological character which have been partially or totally underwater, periodically or continuously, for at least 100 years"; including (i) sites, structures, buildings, artefacts and human remains, together with their archaeological and natural context; (ii) vessels, aircraft, other vehicles or any part thereof, their cargo or other contents, together with their archaeological and natural context; and (iii) objects of prehistoric character [1].

The are many threats to UCH, including treasure hunting, pillage and commercial exploitation, trawling, irresponsible diving ("collecting souvenirs") and unsustainable tourism, resource extraction (sand, gravel, etc.), natural phenomena (earthquakes) and climate change [1]. The last is one of the most alarming issues. Three critical climate-related changes will affect UCH [2]. First, the increase in the surface water temperature will gradually spread to deeper layers, leading to chemical changes, which will provoke the deterioration of UCH. Second, the changes in the current patterns. Some experts predict that climate change could cause a possible interruption of the thermohaline circulation, primarily responsible for regulating the Earth's temperature. Such disruption will modify the sediment layer that preserves most UCH sites/items. Third, the rising

J. M. Machado et al. (Eds.): DCAI 2022, LNNS 585, pp. 127–136, 2023.
https://doi.org/10.1007/978-3-031-23210-7_12

sea level could directly impact not only coastal but also underwater heritage. Some land-based sites will flood. Additionally, extreme weather events (tropical storms, hurricanes, cyclones, etc.) will erode land and underwater heritage [2].

The conservation and restoration of UCH require complete knowledge of the environment in which an item/site is located, the materials from which it is made, and the degradation phenomena experienced in the surrounding environment, which may be physico-chemical (seawater), biological (living organisms), geological (type of substrates and sediments) or human-made [3]. Protection and conservation methods and processes, from the evaluation and analysis of the state of the heritage to restoration activities, still present multiple challenges [4]:

- lack of knowledge and techniques suitable for underwater in situ conservation and protection;
- the elevated costs and the complexity of operating underwater;
- lack of regulation (planning policies, methods, tools, and resources);
- ineffective protection of items/sites and inability to use them for sustainable and responsible tourism development;
- lack of policies and resources to cope with the effects of climate change.

To overcome most of these obstacles, the UNESCO created a treaty, *The Convention on the Protection of Underwater Cultural Heritage 2001* [5], which establishes basic principles for protection, rules for heritage treatment and a system of international cooperation. So far, only 63 countries have ratified or accepted this document.

This work proposes a platform based on the Internet of Underwater Things technologies and the Edge Computing paradigm. It will incorporate Artificial Intelligence techniques that support the monitoring and management of UCH. These algorithms and models, capable of generating knowledge, will work as a supporting tool for decision-making. The rest of the paper is organised as follows: Sect. 2 introduces a sustainable solution for the conservation of UCH, Sect. 3 presents the proposed IoUT-based platform, and Sect. 4 draws the main conclusions and describes future lines of research.

2 A Sustainable Solution for the Conservation of Underwater Cultural Heritage

The conservation of submerged archaeological complexes requires the adoption of innovative and sustainable solutions that aim not only to preserve them in-situ but also to use the available information for decision-making. With the current availability of an enormous amount of data, the challenge is to identify intelligent and adaptive ways of combining the information to create valuable knowledge [6].

The use of sensors could be one of the most cost-effective practices for assessing the state of tangible heritage, facilitating the monitoring of environmental changes. The Internet of Things (IoT) refers to the connection of multiple and heterogeneous objects with electronic devices through different communications to collect and provide data [7–11]. This new technology has grown rapidly, finding applications in multiple sectors such as energy efficiency, health care, industry 4.0, security and public protection logistics and transport, etc. The conservation of UCH can be improved by using efficient and monitoring control systems [12]. On the one hand, preventive conservation is crucial to control the deterioration/decay phenomena of items and sites [13]. To ensure preventive conservation, environmental variables should be long-term monitored and predicted with enough time to react, performing data analytics to detect patterns and dangerous oscillations [14]. On the other hand, long-term monitoring and predictive maintenance can mitigate the damage and reduce future restoration costs [13].

3 An Intelligent Platform for the Management of Underwater Cultural Heritage

Wireless Sensor Network (WSN) plays a key role in IoT. It consists of many distributed sensors interconnected through wireless links for physical and environmental monitoring purposes [15]. The concept of IoT adapted to marine environments is known as the Internet of Underwater Things (IoUT), defined as "the network of smart interconnected underwater objects", such as different types of underwater sensors, autonomous vehicles, buoys, ships, etc. [15, 16]. To support the IoUT, the Underwater Wireless Sensor Networks (UWSN) are considered a promising network system [17, 18]. Since the technologies of communication and waterproofing of equipment are in a mature phase, it is an appropriate time to investigate this field.

Sensors are nodes with acoustic modems distributed in either shallow or deep water. Each sensor node can measure, relay, and forward data. Information is sent through acoustic channels to the components on the surface, called sinks, which are nodes with both acoustic and radio modems [19]. When data arrive at sinks, they will forward it to the remote monitoring centre through radio channels [17]. The sensor nodes, either fixed or mobile, are used to respond to changes in their environments. Physical sensors measure different parameters like temperature, humidity, pressure, etc. Chemical sensors measure parameters like salinity, turbidity, pH, nitrate, chlorophyll, and dissolved oxygen, etc. [20]. Autonomous Underwater Vehicles (AUV) are fitted with sonar sensors for assisted navigation and are equipped with selected sensors that collect data from the surveyed environment [21–24]. Underwater sensors and AUVs cooperate in sophisticated tasks such as large-scale, long-term perception of the environment and reaction to environmental changes [21, 22].

Typically, multiple wireless communication technologies are used in IoUT-based marine environment systems. Underwater acoustic communication technologies are used for data collection and communication among underwater marine environment sensors [25, 26]. Generally, longer-range communication consumes more energy. The selection of the most appropriate wireless communication technology for an application depends on the transmitted data volume, transmission frequency, transmission distance, and available power supply. Nevertheless, significant progress has already been made [15]. Different wireless communication standards and technologies have been developed and tested, including WiFi (range < 100 m), Bluetooth (range 1–100 m), GPRS (dependent on the service provider), ZigBee (range < 75 m) and WiMAX (range < 10 km). A summary and comparison of them can be found in [20].

This work proposes a platform based on IoUT technologies and the Edge Computing paradigm. It will incorporate AI techniques that support the monitoring and management of UCH. These algorithms and models, capable of generating knowledge, will work as a supporting tool for decision-making. The Edge Computing paradigm brings computation and data storage closer to the location where it is required. Its main objective is to solve congestion and bottlenecks in the processing and communication levels, bringing computer resources and services closer to the end-user and the deployed devices [27–30]. The platform will integrate information stored in databases with data acquired in real-time. The proposed platform will work independently and in collaboration with other platforms and systems. IoUT systems part of the proposed platform will be used for:

- monitor and control the heritage's environmental and material conditions to optimally preserve them, detecting structural changes in materials;
- environmental monitoring, including water quality, chemical and biological pollution, thermal pollution, pressure, temperature, pH, salinity, biological growth, conductivity, marine currents, disaster prevention, etc.
- alert of an anomalous event.

Among the main functionalities of the platform, we emphasise the capability to record physical and chemical parameters and transmit the information wirelessly and encrypted. At the sensors level, the proposed architecture will be able to deal with high energy constraints and support fast-changing environments. It will provide elasticity at the storage level and facilitate access to data visualisation.

3.1 A Layered-Based Architecture

Through the platform modules, the control of the different components will be performed by developing a layered architecture. This architecture allows an incremental development of application and service management, and it includes three layers (Fig. 1).

The *perception layer* includes the IoUT sensor and actuator devices, with the objective of sensor data collection and command actuation, surface stations (sinks) and data storage tags [31]. The *network layer* is made of edge nodes, liable for data processing, functioning as data collectors, providing computing, storage, network and other infrastructure resources. This layer allows access of the sink over the sea to the radio channel to process and transmit the information obtained from the perception layer. This information is retransmitted to the onshore centre using different access networking technologies [19, 21]. The *application layer* includes multiple cloud services responsible for data analysis and visualisation, employed for big data analysis and data mining, it is the key where models are performed with AI methodologies from the data uploaded by the edge equipment [11, 21, 32–34].

3.2 Data Stages

The data will combine real-time data collected by the IoUT sensors and information from public databases. The sensors will measure the key environmental factors that participate

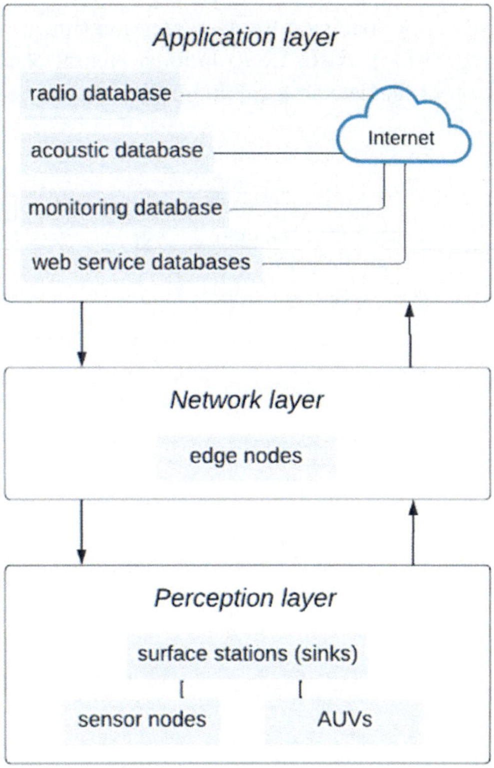

Fig. 1. Proposed architecture. Adapted from [22].

in the degradation phenomena of UCH: temperature, pH, salinity, conductivity, marine currents, biological growth or chemical pollution, especially for those UCH items that are close to the coast. Real-time images will also be collected to show degradation over time and for the monitoring of UCH after the application of the restoration treatments. Information from public databases will include a description of the item (physicochemical properties, age, etc.), the previous restoration works in place, and historical data on key environmental factors.

With this data we aim to:

- describe the current state of UCH, and study the environment in which it is located, so that conservation approaches can be applied;
- make predictions on how degradation phenomena will affect UCH in a changing environment.

The different stages of the data are displayed in Fig. 2. For the pre-processing stage, information will be extracted and transform into a comprehensible structure for further use. Some data mining tools useful for this stage are WEKA, SPSS or KNIME. Later, AI algorithms will be used to quantify UCH degradation phenomena (predictions). The algorithms selected will depend on the data. Several examples of supervised algorithms are classification (decision trees), regression (linear regression), neural networks (non-linear regression). Also, Convolutional Neural Network for images. And lastly, the visualisation of the data so local authorities, public and private organisations and the public can make sustainable decisions regarding UCH.

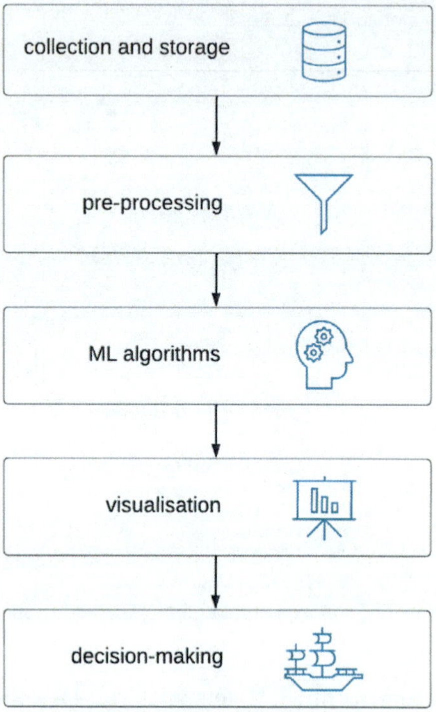

Fig. 2. Proposed data stages.

3.3 Main Challenges

See Table 1.

Table 1. Main challenges of the proposed platform.

Transmision media	UWSNs rely on acoustic communications instead of radio communications because radio signals are absorbed by water quickly. Unfortunately, the communication protocols applied to terrestrial networks cannot be directly applied to acoustic waves [17]
Propagation speed	The propagation speed of UWSN acoustic channels is 1500 m/s [32]; 200,000 times slower than terrestrial networks

(*continued*)

Table 1. *(continued)*

Transmission range	To avoid being absorbed by water, signals need to be transmitted using low frequency. However, lower frequency implies a longer transmission range, with interferences and collisions possibly happening [17]
Transmission rate	Acoustic communications in UWSNs use a narrow bandwidth, so that the transmission rate in UWSNs is generally very low [17]
Difficulty to recharge	Since sensors are deployed underwater, it is challenging and high-priced to recharge their batteries [35]
Energy efficiency	To make the sensor nodes operate for long periods of time, energy efficient algorithms specially designed for IoUT are needed [32, 36]
Mobility	The sensor nodes move and suffer from dynamic network topology changes, impacted by water currents repeatedly. Still, sensor nodes are able to respond to changes in their environments
Locatisation	There are several challenges in complying with the localisation requirements posed by UWSNs [37]. First, the tight time synchronisation between the transmitter and the receiver clocks [38]. Second, the speed of sound cannot be assumed constant (as in the terrestrial localisation schemes); it is a function of temperature, salinity and depth. Third, the multipath effects due to surface reflection, bottom reflection and backscattering [39]
Reliability	The reliability of a link indicates the successful delivery ratio between a pair of sensor nodes. In UWSNs, this ratio would be affected by transmission loss (signals are absorbed by water) and environmental noise. Low reliability leads to frequent data retransmission, resulting in longer delay and higher bandwidth consumption [17, 40]
Lack of standarisation	The development of novel protocols for IoUT is required to provide interoperability between heterogeneous underwater objects [32, 41]
Confidentiality	It refers to protecting information from unauthorised access and preserving the IoUT devices and actuators. Confidentiality is challenging due to the high number of devices involved [42, 43]
Integrity	It refers to data consistency, accuracy, and validity over workflow [44, 45]. In IoUT systems, integrity can safeguard the system against the unapproved spread or modification of information [46]
Availability	It guarantees that service and network will stay operating even in the condition of faults or malicious activities. Availability needs security and a fault management process [47]

4 Conclusions and Future Work

The documentation and conservation of UCH are crucial to preserving society's identity and memory, securing its accessibility to current and future generations. This work proposes a platform based on IoUT technologies and the Edge Computing paradigm. It will incorporate AI techniques that support the monitoring and management of UCH. These algorithms and models, capable of generating knowledge, will work as a supporting tool for decision-making. The platform will integrate information stored in databases with data acquired in real-time. The platform will be able to work independently and/or in collaboration with other existing platforms and systems. Future work includes the implementation and validation of the platform presented in a simulated scenario, where tests can be run.

Acknowledgments. This research has been supported by the project "Technological Consortium TO develop sustainability of underwater Cultural heritage (TECTONIC)", financed by the European Union (Horizon 2020 research and innovation programme under the Marie Skłodowska-Curie grant agreement No 873132).

References

1. UNESCO: Underwater Cultural Heritage. http://www.unesco.org/new/en/culture/themes/underwater-cultural-heritage/. Last accessed: 19 April 2022
2. Pérez-Álvaro, E.: Climate change and underwater cultural heritage: impacts and challenges. J. Cult. Herit. **21**, 842–848 (2016)
3. Memet, J.B.: Conservation of underwater cultural heritage: characteristics and new technologies. Museum Int. **60**, 42–49 (2018)
4. TECTONIC 2022: The Underwater Cultural Heritage: an interdisciplinary challenge. https://www.tectonicproject.eu/about/. Last accessed: 02 May 2022
5. UNESCO: Convention on the Protection of the Underwater Cultural Heritage. https://en.unesco.org/about-us/legal-instruments/convention-protection-underwater-cultural-heritage. Last accessed: 19 April 2022
6. Silva, B.N., Khan, M., Han, K.: Towards sustainable smart cities: a review of trends, architectures, components, and open challenges in smart cities. Sustain. Cities Soc. **38**, 697–713 (2018)
7. Pérez-Pons, M.E., Parra-Domínguez, J., Chamoso, P., Plaza, M., Alonso, R.: Efficiency, profitability and productivity: Technological applications in the agricultural sector. ADCAIJ: Adv. Distrib. Comput. Artif. Intell. J. **9**(4) (2020)
8. Alonso, R.S., Sittón, I., García, O., Prieto, J., Rodríguez-González, S.: An intelligent Edge-IoT platform for monitoring livestock and crops in a dairy farming scenario. Ad Hoc Netw. **98**, 102047 (2020)
9. Márquez-Sánchez, S.: Integral support predictive platform for industry 4.0. ADCAIJ: Adv. Distrib. Comput. Artif. Intell. J. **9**(4), 71–82 (2020)
10. González-Briones, A., Castellanos-Garzón, J., Mezquita Martín, Y., Prieto, J., Corchado, J.: A framework for knowledge discovery from wireless sensor networks in rural environments: a crop irrigation systems case study. Wireless Commun. Mob. Comput. (2018)
11. Aoki, T., Ueno, M.: Photograph classification based on main theme and multiple values by deep neural networks. Adv. Intell. Syst. Comput. **1237** AISC, 206–210 (2021)
12. Plaza-Hernández, M.: An intelligent platform for the monitoring and evaluation of critical marine infrastructures. In: Prieto, J., Pinto, A., Das, A., Ferretti, S. (eds) Blockchain and Applications. BLOCKCHAIN 2020. Advances in Intelligent Systems and Computing, vol. 1238. Springer, Cham (2020)
13. Astorga-González, E.M., Municio, E., Noriega-Alemán, M., Marquez-Barja, J.M.: Cultural heritage and internet of things. In: 6th EAI International Conference on Smart Objects and Technologies for Social Good (2020)

14. Perles, A., et al.: An energy-efficient internet of things (IoT) architecture for preventive, conservation of cultural heritage. Futur. Gener. Comput. Syst. **81**, 566–581 (2018)
15. Xu, G., Shi, Y., Sun, X., Shen, W.: Internet of things in marine environment monitoring: a review. Sensors **19**(7), 1711 (2019)
16. Liou, E., Kao, C., Chang, C., Lin, Y., Huang, C.: Internet of underwater things: Challenges and routing protocols. In: 2018 IEEE International Conference on Applied System Invention (ICASI), pp. 1171–1174 (2018)
17. Kao, C.C., Lin, Y.S., Wu, G.D., Huang, C.J.: A Comprehensive study on the internet of underwater things: applications, challenges, and channel models. Sensors **17**(7), 1477 (2017)
18. Urunov, K., Shin, S., Namgung, J., Park, S.: High-level architectural design of management systems for the internet of underwater things. In: 2018 Tenth International Conference on Ubiquitous and Future Networks (ICUFN) (2018)
19. Qiu, T., Zhao, Z., Zhang, T., Chen, C., Chen, C.L.P.: Underwater internet of things in smart ocean: system architecture and open issues. IEEE Trans. Industr. Inf. **16**, 4297–4307 (2020)
20. Xu, G., Shen, W., Wang, X.: Applications of wireless sensor networks in marine environment monitoring: a survey. Sensors **14**, 16932–16954 (2014)
21. Domingo, M.C.: An overview of the internet of underwater things. J. Netw. Comput. Appl. **35**, 1879–1890 (2012)
22. Krishnaraj, N., Elhoseny, M., Thenmozhi, M., Selim, M.M., Shankar, K.: Deep learning model for real-time image compression on Internet of Underwater Things (IoUT). J. Real-Time Image Proc. pp. 1–15 (2019)
23. Plaza-Hernández, M., Gil-González, A., Rodríguez-González, S., Prieto-Tejedor, J., Corchado-Rodríguez, J.: Integration of IoT technologies in the maritime industry. Adv. Intell. Syst. Comput. **1242** AISC, 107–115 (2021)
24. Plaza-Hernández, M. An iot-based rouv for environmental monitoring. Adv. Intell. Syst. Comput. **1239** AISC, 267–271 (2021)
25. Zixuan, Y., Zhifang, W., Chang, L.: Research on marine environmental monitoring system based on the Internet of Things technology. In: Proceedings of the IEEE International Conference on Electronic Information and Communication Technology (ICEICT) (2016)
26. Jouhari, M., Ibrahimi, K., Tembine, H., Ben-Othman, J.: Underwater wireless sensor networks: a survey on enabling technologies, localization protocols, and internet of underwater things. IEEE Access **7**, 96879–96899 (2019)
27. Alonso, R.S., Sittón, I., Casado-Vara, R., Prieto, J., Corchado, J.M.: Deep reinforcement learning for the management of software-defined networks and network function virtualization in an-Edge-IoT architecture. Sustainability **12**(14), 5706 (2020)
28. Sittón, I., Alonso, R.S., García, O., Muñoz, L., Rodríguez-González, S.: Edge computing, IoT and social computing in smart energy scenarios. Sensors **19**(15), 3353 (2019)
29. Chamoso, P., González-Briones, A., De La Prieta, F., Venyagamoorthy, G., Corchado, J.: Smart city as a distributed platform: toward a system for citizen-oriented management. Comput. Commun. **152**, 323–332 (2020)
30. Alonso, R., Sittón, I., Rodríguez-González, S., García, Ó., Prieto, J.: A survey on software-defined networks and edge computing over IoT. Commun. Comput. Inf. Sci. **1047**, 289–301 (2019)
31. Antao, L., Pinto, R., Reis, J.P., Gonçalves, G.: Requirements for testing and validating the industrial internet of things. In: 11th IEEE conference on software testing, validation and verification (2018)
32. Nayyar, A., Ba, C.H., Coug Duc, N.P., Binh, H.D. Smart-IoUT 1.0: A smart aquatic monitoring network based on internet of underwater things (IoUT). In: Industrial Networks and Intelligent Systems, **257**, 191–207 (2019)

33. Hussain, A., Ullah, I., Hussain, T.: The approach of data mining: a performance-based perspective of segregated data estimation to classify distinction by applying diverse data mining classifiers. ADCAIJ: Adv. Distrib. Comput. Artif. Intell. J. **10**(4), 339–359 (2022)
34. Arora, A., Shoeibi, N., Sati, V., González-Briones, A., Chamoso, P., Corchado, E.: Data augmentation using gaussian mixture model on csv files. Adv. Intell. Syst. Comput. **1237** AISC, 258–265 (2021)
35. Casado-Vara, R., Martin-del Rey, A., Affes, S., Prieto, J., Corchado, J.: IoT network slicing on virtual layers of homogeneous data for improved algorithm operation in smart buildings. Futur. Gener. Comput. Syst. **102**, 965–977 (2020)
36. García, Ó., Alonso, R., Prieto, J., and Corchado, J.: Energy efficiency in public buildings through context-aware social computing. Sensors (Switzerland), **17**(4) (2017)
37. Tan, H.P., Tan; Diamant, R., Seah, W.K.G., Waldmeyer, M.: A survey of techniques and challenges in underwater localisation. Ocean Eng. **38**, 1663–1676 (2011)
38. Casado-Vara, R., Novais, P., Gil, A., Prieto, J., Corchado, J.: Distributed continuous-time fault estimation control for multiple devices in IoT networks. IEEE Access **7**, 11972–11984 (2019)
39. Chandrasekhar, V., Seah, W.K.G., Choo, Y.S., Ee, H.V.: Localization in underwater sensor networks: survey and challenges. In: Proceedings of the 1st ACM International Workshop on Underwater Networks, pp. 33–40 (2006)
40. Assiri, F.: Methods for assessing, predicting, and improving data veracity: a survey. ADCAIJ: Adv. Distrib. Comput. Artif. Intell. J. **9**(4), 5–30 (2020)
41. Salazar, G., Figueiredo, L., Ferreira, N.: Towards the development of iot protocols. Adv. Intell. Syst. Comput. **1239** AISC, 146–155 (2021)
42. Al-, F., Alturjman, S.: Confidential smart-sensing framework in the iot era. J. Supercomput. **74**(10), 5187–5198 (2018)
43. Casado-Vara, R., de la Prieta, F., Prieto, J., Corchado, J.M.: Blockchain framework for IoT data quality via edge computing. In: Proceedings of the 1st Workshop on Blockchain-enabled Networked Sensor Systems, pp. 19–24 (2018)
44. Machado, C., Fröhlich, A.A.M.: Iot data integrity verification for cyberphysical systems using blockchain. In: 2018 IEEE 21st International Symposium on Real-Time Distributed Computing (ISORC). IEEE, pp. 83–90 (2018)
45. González-Briones, A., Chamoso, P. Barriuso, A.: Review of the main security problems with multi-agent systems used in E-commerce applications. ADCAIJ: Adv. Distrib. Comput. Artif. Intell. J. **5**(3), 55–61 (2016)
46. Ahmad, P.: A review on blockchain's applications and implementations. ADCAIJ: Adv. Distrib. Comput. Artif. Intell. J. **10**(2) (2021)
47. Oliveira, P., Pedrosa, T., Novais, P., Matos, P.: Towards to secure an IoT adaptive environment system. Adv. Intell. Syst. Comput. **801**, 349–352 (2019)

Doctoral Consortium

The aim of the Doctoral Consortium Session is to provide a framework as part of which students can present their ongoing research work, meet other students and researchers, and obtain feedback on lines of research for the future.

The Doctoral Consortium is intended for students who have a specific research proposal and some preliminary results, but who are still far from completing their dissertation.

Organizing Committee

Sara Rodríguez, University of Salamanca, Spain

Check for updates

Overview: Security in 5G Wireless Systems

Carlos D. Aguilar-Mora(✉)

Universidad Nacional de Loja, Av. Pío Jaramillo Alvarado, Loja, Ecuador
cdaguilar@usal.es

Abstract. Wireless systems have evolved and are becoming increasingly important because of their capacity and flexibility to support new services and technologies such as enhanced mobile broadband (eMBB), massive machine type communications (mMTC), massive internet of things (mIoT) and ultra-reliable low latency communications (URLLC). This paper presents a systematic review of the status of the fifth-generation or 5G wireless system, addresses its physical architecture as well as the service-based architecture. The main elements of network slicing as a different vision of 5G, in response to the need to split the physical network into multiple logical networks to serve various types of services based on different requirements. Software-defined networking (SDN) and network functions virtualization (NFV) are also presented as tools needed for network resource management. The main security challenges of 5G networks are analyzed and possible solutions are discussed. A comparative analysis is presented of 5G versus its 4G predecessor.

Keywords: 5G · Network Slicing · 5G Security

1 Introduction

Technological evolution and user demand for network services have jointly led to improvements in network resource connectivity and use. Nonetheless, the needs for connection at any geographic location, as well as the variable connection requirements that depend on the intended purpose of the connection, mean that the deployment of communications infrastructure and network traffic will increase substantially. 4G networks are no longer able to support these demands, which poses a great challenge to manufacturers, researchers, professionals and telecommunications agencies. In this context, 5G is the technology that will meet current and future connectivity needs with high-speed rates, shorter response times and the ability to support a large number of connected devices. To offer the new benefits, 5G proposes, in addition to its physical architecture, a different architecture based on services, which consists of separating the control plane from the data plane (or user plane) through the use of software components. Generally speaking, the 5G architecture involves softwarization [1]; which is possible through software-defined networking (SDN), network functions virtualization (NFV) and network slicing [2], as a distinct vision of the 5G network, which involves the division of the physical network into multiple isolated logical networks of different sizes and structures,

J. M. Machado et al. (Eds.): DCAI 2022, LNNS 585, pp. 139–146, 2023.
https://doi.org/10.1007/978-3-031-23210-7_13

that are dedicated to multiple types of services in accordance with their needs, and have different characteristics and requirements [3].

Similarly, although NFV is very important for future communication networks, it has underlying security challenges such as confidentiality, integrity, authenticity and non-repudiation. From the point of view of its use in mobile networks, current NFV platforms do not provide adequate security and isolation for virtualized telecommunication services. One of the main challenges that persist in the use of NFV in mobile networks is the dynamic nature of virtual network functions (VNFs) that leads to configuration errors and therefore security failures. In addition, NFVs are vulnerable to common cyber-attacks such as spoofing, sniffing and DoS. NFV is also vulnerable to a special set of virtualization threats, such as flooding attacks, hypervisor hijacking, malware injection and virtual machine (VM) migration-related attacks, as well as cloud-specific attacks [4].

2 5G Architecture

The 5G network is expected to offer different types of services, including: enhanced mobile broadband (eMBB), massive machine-like communications (mMTC), massive internet of things (mIoT) [5–9] and ultra-reliable, low-latency communications (URLLC); based on different quality of service (QoS) requirements [10]. To achieve the delivery of these multiple services that have different characteristics and needs, it is necessary to identify the improvements to be made to the existing architectures. 5G therefore defines:

2.1 Physical Architecture

The physical architecture of 5G requires the deployment of advanced technologies: dense cells, D2D, CR, cloud, SDN, mm-wave, multi-RAT (Radio Access Technologies), massive-MIMO (Multiple-Input Multiple-Output), leading to more complicated architectures with multiple tiers. Each tier has a different size, transmission power, connections and different RATs. The physical architecture is the basis for the implementation of new architectures and the deployment of services.

2.2 Service-Based Architecture

The service-based architecture consists of separating the control plane from the data plane (or user plane) using software components. This introduces speed and flexibility into the 5G network. Applying SDN solves the problem of configuring and maintaining multiple servers and routers in a dense 5G environment. RANs use SDN to provide intelligence, self-configuration and control plane optimization; in addition, NFV that facilitates 5G network operation to meet user requirements and QoS, network functions are defined: control plane: authentication server function (AUSF), access and mobility management function (AMF), network exposure function (NEF), network repository function (NRF), network slice selection function (NSSF), policy control function (PCF), session management function (SMF), unified data management (UDM), unified data repository (UDR), application function (AF), network data analytics function (NWDAF), charging function (CHF); user plane: user equipment (UE), RAN, user plane function (UPF), data network (DN) [10].

Network Slicing. Network slicing, or 5G network slicing first introduced by the Next Generation Mobile Network (NGMN). As defined by the NGMN, a network slice is a logical network slice running on a common underlying infrastructure (physical or virtual), mutually isolated, with independent control and management that can be created on demand. A network slice can consist of separate domain components in the same or different administrations, or components applicable to the access network, transport network, core network and edge networks. Figure 1 depicts the network slicing scheme.

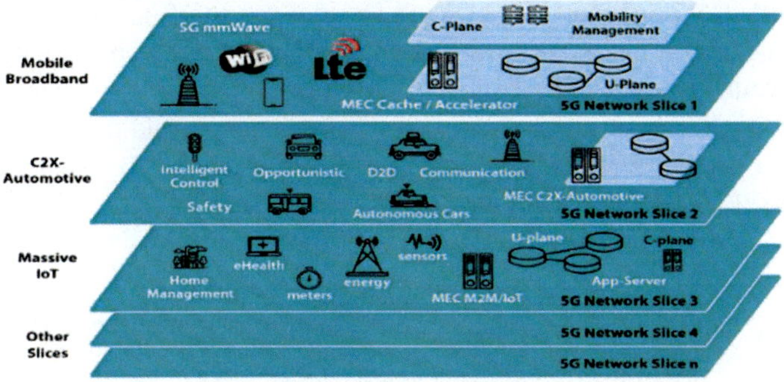

Fig. 1. Network slicing [3]

2.3 SDN

SDN was formally defined by the Open Networking Forum (ONF), a user-driven organization dedicated to the promotion and adoption of SDN through open standards. SDN separates the control functions of networking devices from the data they carry and the switch software from the network hardware. The OpenFlow standard integrates the network control plane in the software running on a server or a network controller, allowing network control to be programmed directly and the underlying infrastructure to be abstracted for applications and network services [11, 12].

2.4 NFV

NFV, is a network architecture that virtualizes entire classes of network node functions into blocks. The goal of NFV is to decouple network functions from dedicated hardware devices and allow network services, now performed by routers, firewalls, load balancers and other dedicated hardware devices, to be hosted on virtual machines (VMs). In this way, operators can design their networks by deploying network services on standard devices; considering that in the 5G era, the network must be able to meet a huge number of diversified service demands from users at different data rates [11]. The virtualization of network functions requires a management and orchestration system (MANO) to manage the virtualized infrastructure, cloud systems, communication and network infrastructure, software-defined networks, NFV entities and the different lifecycles of all these components [13, 14].

3 Security in 5G Networks

Communications over 5G networks are currently in the spotlight of industry, academia and governments; intruders or attackers are also vigilant and surely working on networks that pave the way for new attacks. 5G uses several relatively new technologies including software-defined networking (SDN) and network functions virtualization (NFV) to meet the many new requirements for different network capabilities. Its ability to support a large number of connected devices and the innovation of new techniques mean that the new cyber threat landscape will be dynamic and complex, posing significant challenges to security, privacy and confidentiality [15]. Security in 5G can be categorized into availability, authentication, non-repudiation, integrity and confidentiality. Several technologies used in 5G may be compromised: Industry 4.0 [16, 17], Blockchain [9, 18–22], AI [23, 24], D2D communication, Mobile Edge Computing [25–28], SDN [29].

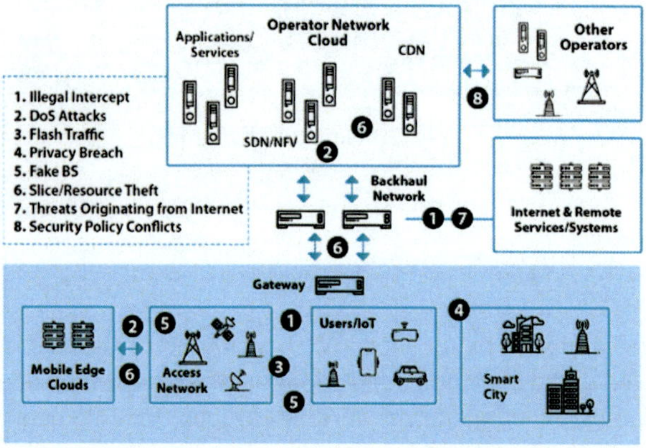

Fig. 2. Threat landscape in a 5G scenario [30]

Generally, most networks face the following types of attacks [31]: DDoS (Distributed Denial of Service), IP address spoofing, flow table overloading, control plane saturation, host location hijacking, fragmentation, SYN flood. In 5G networks, the following attacks may occur: illegal interception, DoS, flash traffic, privacy violation, false base station, resource/slice theft, threats originating from the Internet, conflict of security policies. Figure 2 shows the possible attacks in a 5G scenario.

However, according to [32] network virtualization implies that other security aspects must also be addressed, such as: identity and access management, API security, network anomaly and intrusion detection and prediction, root cause analysis, moving target defense (MTD). Therefore, secure network architectures, mechanisms and protocols are needed as a basis for the 5G network to address this problem, following security standards in both design and operations.

4 5G Versus 4G

According to [10], 5G networks meet several requirements compared to its 4G predecessor: They support a massive number of connected devices, anywhere from 10 to 100 times more than 4G. They support a volume of mobile data per area 1000 times higher than 4G. They provide 10–100 times higher data rate than 4G. 5 times the End-to-End (E2E) latency, reaching 5 ms. They provide nearly 100% availability and 100% geographical coverage. They support the coexistence of different radio access network (RAN) technologies. Increased security and privacy. Their power consumption is 10 times lower than 4G. 5G supports real-time processing and transmission. Network management costs are 5 times lower than 4G. They provide a seamless integration of current wireless technologies. The network will be more flexible, intelligent, dynamic and open service. These networks are more cost-effective in terms of capital and operating expenditures (CAPEX and OPEX). The battery life of devices is 10 times longer than 4G.

5 Potential Solutions to Security Problems

With the arrival of 5G, standards are especially important on a global scale. Typically, a standard is the key to the convergence of the telecommunications and IT sectors to develop a ubiquitous infrastructure. This ubiquitous infrastructure offers global services to customers and creates new opportunities to interconnect a wide range of smart objects. To ensure the full benefits of 5G, all security events or issues that accompany the 5G architecture must be dealt with in a standardized manner [33].

As even more user data and network traffic will be transferred in 5G networks, big data security solutions supported by artificial intelligence (AI) techniques, must be sought to cope with the magnitude of the volume of damage and ensure the security issues at stake.

The adoption of AI to address different threats is a great option, thanks to its potential to empower intelligent, adaptive and autonomous security management [34, 35].

6 Conclusion

The fifth-generation wireless system, also called 5G, presents a promising scenario for different services and applications. However, it gives rise to a number of security threats that can compromise the infrastructure supporting 5G. In this paper we have outlined the main features of 5G, its architecture, the security issues to be addressed, as well as possible solutions to the different problems; a list of key features in 5G related elements is included, it is based on the available literature. SDN/NFV and network slicing features are being designed and developed with a focus on delivering on the requirements of flexibility, agility and scalability, taking advantage of recent advances in virtualized services, AI and cloud computing.

To address the problem of security in 5G networks, the main lines of research to be considered in the development of the research project focus on the analysis of secure architectures for 5G networks and the study of artificial intelligence techniques for network slicing.

Acknowledgment. This work has been partially supported by the Institute for Business Competitiveness of Castilla y León, and the European Regional Development Fund under grant CCTT3/20/SA/0002 (AIR-SCity project).

References

1. Batalla, J.M., et al.: Security risk assessment for 5G networks: national perspective. IEEE Wirel. Commun. **27**, 16–22 (2020). https://doi.org/10.1109/MWC.001.1900524
2. Casado-Vara, R., Martin-del Rey, A., Affes, S., Prieto, J., Corchado, J.M.: IoT network slicing on virtual layers of homogeneous data for improved algorithm operation in smart buildings. Future Gener. Comput. Syst. **102**, 965–977 (2020). https://doi.org/10.1016/j.future.2019.09.042
3. Barakabitze, A.A., Ahmad, A., Mijumbi, R., Hines, A.: 5G network slicing using SDN and NFV: a survey of taxonomy, architectures and future challenges. Comput. Netw. **167**, 106984 (2020). https://doi.org/10.1016/j.comnet.2019.106984
4. Ahmad, I., Kumar, T., Liyanage, M., Okwuibe, J., Ylianttila, M., Gurtov, A.: Overview of 5G security challenges and solutions. IEEE Commun. Stand Mag. **2**, 36–43 (2018)
5. Pérez-Pons, M.E., Parra-Domínguez, J., Chamoso, P., Plaza, M., Alonso, R.: Efficiency, profitability and productivity: technological applications in the agricultural sector. ADCAIJ Adv. Distrib. Comput. Artif. Intell. **J9** (2020). https://doi.org/10.14201/ADCAIJ2020944754
6. Nono, R., Alsudais, R., Alshmrani, R., Alamoudi, S., Aljahdali, A.O.: Intelligent traffic light for ambulance clearance. ADCAIJ Adv. Distrib. Comput. Artif. Intell. **J9**, 89–104 (2020). https://doi.org/10.14201/ADCAIJ20209389104
7. González-Briones, A., Castellanos-Garzón, J.A., Mezquita Martín, Y., Prieto, J., Corchado, J.M.: A framework for knowledge discovery from wireless sensor networks in rural environments: a crop irrigation systems case study. Wirel. Commun. Mob. Comput. **2018**, 1–14 (2018). https://doi.org/10.1155/2018/6089280
8. Valdeolmillos, D., Mezquita, Y., Ludeiro, A.R.: Sensing as a service: an architecture proposal for big data environments in smart cities. In: Novais, P., Lloret, J., Chamoso, P., Carneiro, D., Navarro, E., Omatu, S. (eds.) ISAmI 2019. AISC, vol. 1006, pp. 97–104. Springer, Cham (2020). https://doi.org/10.1007/978-3-030-24097-4_12
9. Mezquita, Y.: Internet of Things platforms based on blockchain technology: a literature review. In: Herrera-Viedma, E., Vale, Z., Nielsen, P., Martin Del Rey, A., Casado Vara, R. (eds.) DCAI 2019. AISC, vol. 1004, pp. 205–208. Springer, Cham (2020). https://doi.org/10.1007/978-3-030-23946-6_26
10. Fourati, H., Maaloul, R., Chaari, L.: A survey of 5G network systems: challenges and machine learning approaches. Int. J. Mach. Learn. Cybern. **12**(2), 385–431 (2020). https://doi.org/10.1007/s13042-020-01178-4
11. Bo, R.: 5G Heterogeneous Networks Self-organizing and Optimization, 1st edn. Springer International Publishing, Cham (2016)
12. Llorens-Carrodeguas, A., Cervelló-Pastor, C., Leyva-Pupo, I.: Software defined networks and data distribution service as key features for the 5G control plane. In: Rodríguez, S., et al. (eds.) DCAI 2018. AISC, vol. 801, pp. 357–360. Springer, Cham (2019). https://doi.org/10.1007/978-3-319-99608-0_45
13. Yousaf, F.Z., Bredel, M., Schaller, S., Schneider, F.: NFV and SDN—key technology enablers for 5G networks. IEEE J. Sel. Areas Commun. **35**, 2468–2478 (2017). https://doi.org/10.1109/JSAC.2017.2760418

14. Leyva-Pupo, I., Cervelló-Pastor, C., Llorens-Carrodeguas, A.: The resources placement problem in a 5G hierarchical SDN control plane. In: Rodríguez, S., et al. (eds.) DCAI 2018. AISC, vol. 801, pp. 370–373. Springer, Cham (2019). https://doi.org/10.1007/978-3-319-99608-0_48
15. Mazurczyk, W., Bisson, P., Jover, R.P., Nakao, K., Cabaj, K.: Special issue on advancements in 5G networks security. Future Gener. Comput. Syst. **110**, 314–316 (2020). https://doi.org/10.1016/j.future.2020.04.043
16. Márquez Sánchez, S.: Integral support predictive platform for industry 4.0. ADCAIJ Adv. Distrib. Comput. Artif. Intell. **J9**, 71–82 (2020). https://doi.org/10.14201/ADCAIJ2020947182
17. Chamoso, P., González-Briones, A., De La Prieta, F., Venyagamoorthy, G.K., Corchado, J.M.: Smart city as a distributed platform: toward a system for citizen-oriented management. Comput. Commun. **152**, 323–332 (2020). https://doi.org/10.1016/j.comcom.2020.01.059
18. Ahmad, P.: A review on blockchain's applications and implementations. ADCAIJ Adv. Distrib. Comput. Artif. Intell. **J10** (2021). https://doi.org/10.14201/ADCAIJ2021102197208
19. Parra Domínguez, J., Roseiro, P.: Blockchain: a brief review of agri-food supply chain solutions and opportunities. ADCAIJ Adv. Distrib. Comput. Artif. Intell. **J9**, 95–106 (2020). https://doi.org/10.14201/ADCAIJ20209495106
20. Rathore, S., Park, J.H., Chang, H.: Deep Learning and blockchain-empowered security framework for intelligent 5G-enabled IoT. IEEE Access **9**, 90075–90083 (2021). https://doi.org/10.1109/ACCESS.2021.3077069
21. Srivastav, R.K., Agrawal, D., Shrivastava, A.: A survey on vulnerabilities and performance evaluation criteria in blockchain technology. ADCAIJ Adv. Distrib. Comput. Artif. Intell. **J9**, 91–105 (2020). https://doi.org/10.14201/ADCAIJ20209291105
22. Casado-Vara, R., de la Prieta, F., Prieto, J., Corchado, J.M.: Blockchain framework for IoT data quality via edge computing. In: Proceedings of the 1st Workshop on Blockchain-enabled Networked Sensor Systems. ACM, Shenzhen China, pp. 19–24 (2018)
23. Marais, B., Quertier, T., Chesneau, C.: Malware Analysis with Artificial Intelligence and a Particular Attention on Results Interpretability (2021)
24. Alonso, R.S.: Deep symbolic learning and semantics for an explainable and ethical artificial intelligence. In: Novais, P., Vercelli, G., Larriba-Pey, J.L., Herrera, F., Chamoso, P. (eds.) ISAmI 2020. AISC, vol. 1239, pp. 272–278. Springer, Cham (2021). https://doi.org/10.1007/978-3-030-58356-9_30
25. Sittón-Candanedo, I.: A new approach: edge computing and blockchain for industry 4.0. In: Herrera-Viedma, E., Vale, Z., Nielsen, P., Martin Del Rey, A., Casado Vara, R. (eds.) DCAI 2019. AISC, vol. 1004, pp. 201–204. Springer, Cham (2020). https://doi.org/10.1007/978-3-030-23946-6_25
26. Sittón-Candanedo, I.: Edge computing: a review of application scenarios. In: Herrera-Viedma, E., Vale, Z., Nielsen, P., Martin Del Rey, A., Casado Vara, R. (eds.) DCAI 2019. AISC, vol. 1004, pp. 197–200. Springer, Cham (2020). https://doi.org/10.1007/978-3-030-23946-6_24
27. Chen, W.-H., Kuo, H.-Y., Lin, Y.-C., Tsai, C.-H.: A lightweight pedestrian detection model for edge computing systems. In: Dong, Y., Herrera-Viedma, E., Matsui, K., Omatsu, S., González Briones, A., Rodríguez González, S. (eds.) DCAI 2020. AISC, vol. 1237, pp. 102–112. Springer, Cham (2021). https://doi.org/10.1007/978-3-030-53036-5_11
28. Plaza-Hernández, M., Gil-González, A.B., Rodríguez-González, S., Prieto-Tejedor, J., Corchado-Rodríguez, J.M.: Integration of IoT technologies in the maritime industry. In: Rodríguez González, S., et al. (eds.) DCAI 2020. AISC, vol. 1242, pp. 107–115. Springer, Cham (2021). https://doi.org/10.1007/978-3-030-53829-3_10
29. Park, J.H., Rathore, S., Singh, S.K., Salim, M.M., Azzaoui, A.E., Kim, T.W., Pan, Y., Park, J.H.: A Comprehensive Survey on Core Technologies and Services for 5G Security: Taxonomies, Issues, and Solutions, p. 23 (2021)

30. Ahmad, I., Shahabuddin, S., Kumar, T., Okwuibe, J., Gurtov, A., Ylianttila, M.: Security for 5G and Beyond **21**, 41 (2019)
31. Abdulqadder, I.H., Zhou, S., Zou, D., Aziz, I.T., Akber, S.M.A.: Multi-layered intrusion detection and prevention in the SDN/NFV enabled cloud of 5G networks using AI-based defense mechanisms. Comput. Netw. **179**, 107364 (2020). https://doi.org/10.1016/j.comnet.2020.107364
32. Benzaïd, C., Taleb, T.: AI for beyond 5G networks: a cyber-security defense or offense enabler? IEEE Netw. **34**, 140–147 (2020). https://doi.org/10.1109/MNET.011.2000088
33. Khan, R., Kumar, P., Jayakody, D.N.K., Liyanage, M.: A survey on security and privacy of 5G technologies: potential solutions, recent advancements, and future directions. IEEE Commun. Surv. Tutor **22**, 196–248 (2020). https://doi.org/10.1109/COMST.2019.2933899
34. Benzaid, C., Taleb, T.: AI for beyond 5G networks: a cyber-security defense or offense enabler? IEEE Netw. **34**, 140–147 (2020). https://doi.org/10.1109/MNET.011.2000088
35. Yigitcanlar, T., et al.: Artificial intelligence technologies and related urban planning and development concepts: how are they perceived and utilized in Australia? J. Open Innov. Technol. Mark Complex **6**, 187 (2020). https://doi.org/10.3390/joitmc6040187

A Study on the Application of Protein Language Models in the Analysis of Membrane Proteins

Hamed Ghazikhani[1](✉) and Gregory Butler[1,2]

[1] Department of Computer Science and Software Engineering, Concordia University, Montreal, Canada
{hamed.ghazikhani,gregory.butler}@concordia.ca
[2] Centre for Structural and Functional Genomics, Concordia University, Montreal, Canada

Abstract. Membrane proteins make up around 30% of all proteins in a cell. These proteins are difficult to evaluate due to their hydrophobic surface and dependence on their original *in vivo* environment. There is a tremendous demand for computational approaches for predicting membrane proteins' function and structure. Protein sequences, the language of life, can benefit from using natural language algorithms. In this study, we propose to use the BERT language model to represent membrane proteins. Our preliminary results, which utilized a Logistic Regression classifier, reveal that BERT can comprehend the context of membrane proteins and enable us to investigate better and distinguish these proteins.

Keywords: Membrane proteins · BERT · Language model · Machine learning · Neural network

1 Problem Statement

Membrane proteins regulate cellular functions such as cell signaling, trafficking, metabolism, and energy generation. Transporters are membrane proteins that govern chemical movement into and out of cells. They are essential for cellular homeostasis and thus attractive targets for pharmaceutical companies [1]. Membrane protein detection and understanding of transporters and processes are critical to progress in functional and structural genomics.

Due to the vast labor required to characterize membrane proteins, their structure and function remain poorly defined and understood [8]. Because the function of a protein may be connected to the native environment in which an organism lives, identifying its function in a laboratory is not an easy task. Thus, it is desirable to use membrane protein sequences with experimental data in computational methods to detect and study them.

While current efforts to annotate membrane proteins are far from ideal, tentative results exist that need to be refined. As in [1,2], state-of-the-art approaches

J. M. Machado et al. (Eds.): DCAI 2022, LNNS 585, pp. 147–152, 2023.
https://doi.org/10.1007/978-3-031-23210-7_14

often characterize membrane proteins with complex methods and feature sets, resulting in intricate algorithms and difficulty in comprehending the results effectively. Furthermore, many other methods (such as [4,14]) rely solely on evolutionary information, which is not available for all proteins (*de novo* or unannotated, dubbed the *dark proteome* [5]), or the model cannot generalize unseen protein sequences because the training and test sets are not independent such as work in [15]. Membrane protein sequence study is complex and time-consuming, so we should propose a simple and reproducible method.

2 Related Work

Significant research has been conducted on studying and predicting membrane proteins, which is addressed in detail in Butts et al. [8] paper. Typically, projects function on two levels: the first identifies if a particular protein sequence is a membrane protein, and the second predicts membrane protein subtypes.

Chou and Shen first suggested MemType-2L in 2007 [9] which makes use of a two-layer predictor; the first layer predicts membrane proteins, while the second layer predicts subtypes. They combined OET-KNN (Optimized Evidence-Theoretic K-Nearest Neighbor) [10] classifiers with Pse-PSSM evolutionary features using an ensemble approach. The first layer correctly predicts membrane proteins 92.7% of the time. Butts et al. [7] proposed a technique for identifying the membrane protein type. They used statistical moments to extract features from protein sequences and trained a feed-forward neural network to predict. When the jackknife test is utilized, the results are 91.23%. Arif et al. (2018) developed iMem-2LSAAC [3], a two-layer approach for predicting membrane proteins and membrane protein subclasses. They accomplished this using a split amino acid composition and a support vector machine. Their approach starts with assessing whether a protein sequence is a membrane protein, and then the second part represents the membrane protein subtype. They exhibit a 94.61% accuracy for membrane protein prediction (first layer) using the jackknife test on their dataset.

Alballa and Butler proposed TooT-M [2] in 2020, an integrative technique for membrane protein prediction. Compositional features, as well as transmembrane topology prediction and evolutionary features, were evaluated in their work. Additionally, they assessed an ensemble learning approach utilizing 50 Support Vector Machine (SVM) classifiers, 50 Gradient Boosting Machine (GBM) classifiers, 50 Random Forest (RF) classifiers, 50 and 500 K-Nearest Neighbor (KNN) classifiers, and 50 and 500 OET-KNN classifiers. Their article [2] compares and contrasts the performance of all feature sets and classifiers. They report an MCC of 0.85 for membrane proteins and prediction accuracy of 92.51%. While integrating several feature sets and classifiers enhances TooT-M's characterization of membrane proteins, it also increases the task's complexity, resulting in a complicated structure. Additionally, these complex procedures prohibit any interpretation or explanation of the methods used to produce these findings.

3 Hypothesis

We can apply natural language approaches since proteins are denoted by string concatenation, like human languages. Modern deep neural network designs for sequences, such as BERT (Bidirectional Encoder Representations from Transformers) [11], have ushered in a revolution in automated text analysis. The Transformer architecture [16] has achieved astounding performance in a variety of benchmarks and applications [6].

Deep language models, such as BERT, are constructed by stacking multi-layers of Transformers encoder that incorporates an attention mechanism capable of determining the (remote) relationship between each amino acid in a sequence. Furthermore, the pre-trained deep language models can be used without requiring high-performance computers due to the concept of *transfer learning*.

From a BERT model, membrane proteins may be represented in two unique ways: frozen and fine-tuned. The former is used to extract features from a pre-trained BERT model without updating its weights, while the latter is used to extract features after training the pre-trained BERT model on a smaller dataset and fine-tuning its weights.

We hypothesize that deep language models may assist in our understanding of membrane proteins. This proposal provides a research hypothesis for overcoming membrane protein analysis challenges using the BERT language model. The following is the formulation of our study question: ***Q1****: Is BERT capable of comprehending membrane protein context?* and ***Q2****: Which BERT representation should be used for membrane protein analysis?* For this, we will examine the following case study: Is the sequence of a protein X a membrane protein or not?

4 Proposal

The pre-trained language models in the ProtTrans project [12] allow us to use transformer-based models. In this study, we propose using BERT (ProtBert-BFD) model from ProtTrans, which is pre-trained on the BFD database to discriminate membrane proteins from non-membrane proteins using a Logistic Regression classifier with default hyperparameters from scikit-learn (https://scikit-learn.org). To this end, we utilized the same dataset as the TooT-M project [2], which is accessible at (https://tootsuite.encs.concordia.ca/datasets/membrane.) The detailed process of collecting and benchmarking the dataset is given in [2] and the total number of sequences in this dataset is shown in Table 1.

We evaluated the proposed technique using a 10-fold Cross-Validation (CV) and Leave-One-Out (LOOCV) along with Sensitivity, Specificity, Accuracy, and MCC (Matthew's Correlation Coefficient) assessment metrics, which are detailed and formulated in [13].

Table 1. Membrane dataset

Class	Training	Test	Total
Membrane	7,945	883	8,828
Nonmembrane	8,157	907	9,064
Total	16,102	1,790	17,892

5 Preliminary Results

We fine-tuned the ProtBert-BFD on membrane dataset (Table 1) to address the following research question: ***Q2****: Which BERT representation should be used for membrane protein analysis?* The results of this experiment are summarized in Table 2, which compares frozen and fine-tuned ProtBert-BFD on an independent test set. According to this table, fine-tuned representations outperform frozen representations.

Table 2. ProtBert-BFD representation comparison

Rep	Sen(%)	Spc(%)	Acc(%)	MCC
Frozen	91.18	83.47	87.37	0.7492
Fine-tuned	91.28	93.61	92.46	0.8493

This table shows the sensitivity, specificity, accuracy, and MCC of the frozen/fine-tuned representations from the ProtBert-BFD model on the independent test set.

The results of combining a Logistic Regression with a fine-tuned ProtBert-BFD representation are displayed in Table 3. This table demonstrates that the LOOCV results provide no more information beyond the CV results.

Table 3. CV, LOOCV and independent test set results

Eval	Sen(%)	Spc(%)	Acc(%)	MCC
CV	98.19	98.74	98.47	0.97
LOOCV	98.14	98.68	98.41	0.97
Independent test	91.28	93.61	92.46	0.85

This table shows the sensitivity, specificity, accuracy, and MCC of the 10-fold CV and LOOCV and an independent test set of fine-tuned representations from the ProtBert-BFD model.

Using the same dataset, the proposed method's performance is compared to state-of-the-art techniques for membrane protein prediction in Table 4.

According to this table, the proposed method outperforms iMem-2LSAAC and MemType-2L across all evaluation metrics. It outperforms TooT-M in terms of specificity while underperforming sensitivity and matching TooT-M for accuracy and MCC.

Table 4. Comparison with other methods

Method	Sen(%)	Spc(%)	Acc(%)	MCC
iMem-2LSAAC [3]	74.52	83.90	79.27	0.59
MemType-2L [9]	88.67	90.19	89.44	0.79
TooT-M [2]	**92.41**	92.5	**92.46**	**0.85**
Proposed Method	91.28	**93.61**	**92.46**	**0.85**

This table compares the performance of previous approaches on the CV and independent test set using sensitivity, specificity, accuracy, and MCC evaluation. The results are reported from [1]. Each column's maximum value is shown in boldface.

Table 5 shows the performance of the proposed method to address our first research question: ***Q1**: Is BERT capable of comprehending membrane protein context?* According to this table, BERT understood the context of the membrane protein, and we achieved an MCC of 0.85 in discriminating membrane proteins from non-membrane proteins.

Table 5. The proposed method performance

Class	TP	TN	FP	FN	MCC
Membrane	806	849	58	77	0.8493

This table presents Logistic Regression performance on the independent test set using a fine-tuned ProtBert-BFD representation model in terms of true positive, true negative, false positive, false negative, and MCC.

6 Reflections

Even though fine-tuned representations outperform frozen representations (Table 2), there is a catch: Is fine-tuning a BERT model always the best option? Updating a BERT model's 420 million parameters takes roughly three days in our case. We then utilized a Logistic Regression with default hyperparameters to investigate BERT representations on membrane proteins. The following issue arose: Do SVM or Convolutional Neural Networks perform better than Logistic Regression? We are certain that using the BERT representation would considerably enhance the field of protein research, particularly membrane proteins.

References

1. Alballa, M.: Predicting Transporter Proteins and Their Substrate Specificity. PhD Thesis, Concordia University (2020)
2. Alballa, M., Butler, G.: Integrative approach for detecting membrane proteins. BMC Bioinform. **21**(19), 575 (2020)
3. Arif, M., Hayat, M., Jan, Z.: iMem-2LSAAC: a two-level model for discrimination of membrane proteins and their types by extending the notion of SAAC into Chou's pseudo amino acid composition. J. Theor. Biol. **442**, 11–21 (2018)
4. Barghash, A., Helms, V.: Transferring functional annotations of membrane transporters on the basis of sequence similarity and sequence motifs. BMC Bioinform. **14**(1), 343 (2013)
5. Bitard-Feildel, T., Callebaut, I.: Exploring the dark foldable proteome by considering hydrophobic amino acids topology. Sci. Rep. **7**(1), 41425 (2017)
6. Brandes, N., Ofer, D., Peleg, Y., Rappoport, N., Linial, M.: ProteinBERT: a universal deep-learning model of protein sequence and function. Bioinformatics (2022)
7. Butt, A.H., Khan, S.A., Jamil, H., Rasool, N., Khan, Y.D.: A prediction model for membrane proteins using moments based features. In: BioMed Research International 2016, e8370132, Hindawi (2016)
8. Butt, A.H., Rasool, N., Khan, Y.D.: A treatise to computational approaches towards prediction of membrane protein and its subtypes. J. Membr. Biol. **250**(1), 55–76 (2017)
9. Chou, K.C., Shen, H.B.: MemType-2L: a web server for predicting membrane proteins and their types by incorporating evolution information through Pse-PSSM. Biochem. Biophys. Res. Commun. **360**(2), 339–345 (2007)
10. Denoeux, T.: A k-nearest neighbor classification rule based on Dempster-Shafer theory. IEEE Trans. Syst. Man Cybern. **25**(5), 804–813 (1995)
11. Devlin, J., Chang, M.W., Lee, K., Toutanova, K.: BERT: Pre-training of Deep Bidirectional Transformers for Language Understanding. arXiv:1810.04805 [cs] (2019)
12. Elnaggar, A., Heinzinger, M., Dallago, C., Rehawi, G., Wang, Y., Jones, L., Gibbs, T., Feher, T., Angerer, C., Steinegger, M., Bhowmik, D., Rost, B.: ProtTrans: towards cracking the language of lifes code through self-supervised deep learning and high performance computing. IEEE Trans. Pattern Anal. Mach. Intell. 1–1 (2021)
13. Grandini, M., Bagli, E., Visani, G.: Metrics for Multi-class Classification: An Overview. arXiv:2008.05756 [cs, stat] (2020). arXiv: 2008.05756
14. Kabir, M., Arif, M., Ali, F., Ahmad, S., Swati, Z.N.K., Yu, D.J.: Prediction of membrane protein types by exploring local discriminative information from evolutionary profiles. Anal. Biochem. **564–565**, 123–132 (2019)
15. Liu, L.X., Li, M.L., Tan, F.Y., Lu, M.C., Wang, K.L., Guo, Y.Z., Wen, Z.N., Jiang, L.: Local sequence information-based support vector machine to classify voltage-gated potassium channels. Acta Biochimica et Biophysica Sinica **38**(6), 363–371 (2006)
16. Vaswani, A., Shazeer, N., Parmar, N., Uszkoreit, J., Jones, L., Gomez, A.N., Kaiser, L., Polosukhin, I.: Attention Is All You Need. arXiv (2017)

Visualization for Infection Analysis and Decision Support in Hospitals

Denisse Kim[1(✉)], Jose M. Juarez[1], Manuel Campos[1,2], and Bernardo Canovas-Segura[1]

[1] AIKE research group (INTICO), University of Murcia, Campus Espinardo, Murcia 30100, Spain
denissemisoo.kim@um.es

[2] Murcian Bio-Health Institute (IMIB-Arrixaca), El Palmar, Murcia 30120, Spain

Abstract. Multridrug-resistant (MDR) bacteria are currently a serious threat to public health. They are primarily associated with hospital-acquired infections and their spread is related to an increase in the morbidity, mortality and healthcare cost. Therefore, their control and prevention is a priority problem today. In this control, the need of detecting MDR-bacteria outbreaks inside a hospital stands out. This complex process requires interweaving reports with temporal and spatial information. In this matter, computer-aided visualization techniques might play an important role, by helping explain and understand data-driven decision making. Thus, the hypothesis of this PhD thesis is that spatial-temporal modeling and visualization techniques allow clinicians to increase their confidence and comprehension of AI-based epidemiological analysis and prediction models. During the first phase of this PhD project, we have carried out a detailed analysis of what has been done on the application of spatial-temporal visualization techniques on epidemiological data. The results of this investigation have helped us identify the current trends and gaps in this field. Following this, we have implemented a simulation model and its simulator, with the objective of generating clinically realistic spatial-temporal data of infection outbreaks within hospitals.

Keywords: Artificial Intelligence · Visualization techniques · Epidemiological analysis

1 Medical Problem and Background

Antimicrobials are medicines used for the treatment and prevention of infections; this includes antibiotics, antivirals, antifungals and antiparasitics. Antimicrobial resistance occurs when bacteria, viruses or others change over time and stop responding to said medicines, making the treatment difficult and increasing the risk of spread, severity of diseases and death [14]. Specifically, multidrug-resistant (MDR) bacteria are one of the critical threats to public health at the moment, as their spread is associated with an increase in the morbidity, mortality and healthcare costs [7].

J. M. Machado et al. (Eds.): DCAI 2022, LNNS 585, pp. 153–158, 2023.
https://doi.org/10.1007/978-3-031-23210-7_15

MDR bacteria are primarily associated with hospital-acquired infections: in a hospital, a patient's exposure or infection with an MDR bacteria can occur through contact with a carrier, exposure to contaminated environments, use of contaminated medical equipment or following the use of antimicrobial agents [13]. The control and prevention of infections caused by MDR bacteria is, therefore, a priority problem today. This control is divided in a series of relevant tasks, which are: early detection of infection outbreaks, communication to healthcare staff and monitoring of cases within a hospital to help prevent the spread [4]. However, the detection of MDR-bacteria outbreaks inside a hospital is a complex process that requires interweaving reports with temporal and spatial information. Computer-aided visualization techniques might play an important role to explain and understand data-driven decision making in the following years.

In the most recent literature, there is growing interest on the computer-based detection and notification of infections in hospital settings. We highlight the following:

- Myal et al. [11] presented a network-based analysis of movement of patients carrying drug-resistant bacteria across several hospitals so as to detect disease transmissions.
- Baumgartl et al. [3] presented a visual analytic approach to support the analysis of contacts between patients, transmission pathways, progression of the outbreak and patient timelines during hospitalization.
- Arantes et al. [2] developed a system for monitoring occurrence trends of hospital-acquired infections using statistical process control charts.

In most cases this research focuses on the monitoring of specific infection scores, the detection of abnormal events and the development of tailored statistical models from a classical approach [1]. Some of them also focus on bacteria's features and contact network models, but neglect the spatial characteristics of hospital buildings.

We have also identified that there is a lack of general reproducibility of the experiments due to (1) the absence of open-access data of this nature; (2) the clinical datasets from hospitals used are confidential, under the personal data regulations. From the scientific point of view, this fact could lead into two problems: (a) the difficulty to establish a fair analysis between previous and current computational models; and (b) the limited use of Machine Learning and Deep Learning methods for building, evaluating their performance and robustness, and their clinical validation.

Therefore, there is a need for new tools able to generate, analyze and visualize spatial-temporal data for MDR-bacteria infections to support decision-making by health personnel and to ease their work.

2 Hypothesis and Planned Approach

Considering all the factors mentioned, the research hypothesis of this PhD thesis is that spatial-temporal modeling and visualization techniques allow clinicians to increase their confidence and comprehension of AI-based epidemiological analysis and prediction models.

This PhD thesis aims to:

- Identify current and most suitable visualization methods for explaining results in epidemiological analysis considering spatial and temporal dimensions.
- Get high quality spatial-temporal dataset of hospital MDR-bacterial infection at our disposal. We will consider both approaches: real datasets from hospital settings and realistic simulated data.
- Propose and design new visualization methods for explaining epidemiological indicators (e.g. incidence, prevalence, mortality) considering spatial and temporal dimensions. We plan to implement an interactive visual tool for the detection of outbreaks and endemics using traditional clinical approaches.
- Develop new ML-based prediction models for hospital outbreak analysis and an extension of the interactive visual tool based on explainable and trustworthy principles.

3 Preliminary Results

During the first phase of this PhD project, the following results have been obtained:

3.1 Visualization Models for Epidemiology Analysis

We have carried out a detailed analysis of what has been done in the last years on the application of spatial-temporal visualization techniques on epidemiological data. To do this, we have conducted a systematic review following the PRISMA (Preferred Reporting Items for Systematic Reviews and Meta-Analyses) guidelines [10] and we have searched for papers in various scientific databases. Following this, we have reviewed 1049 articles, of which 71 were deemed suitable for inclusion in the study. With this systematic review, we have put the efforts made during the last two decades into perspective, paying special attention to how epidemic measures, temporal and spatial information are displayed. We identified that there is an increasing development of visualization tools, since the need to display the data for a better and more efficient interpretation is more valued. We have inferred the current trends in the development of visualization programs and identified some gaps concerning the lack of a standard evaluation methodology, the study of individual epidemiological data and, more specifically, of diseases acquired by patients in hospitals as the result of an epidemic. This research is currently under evaluation at [8].

3.2 Agent-based Simulation of Clostridium Difficile Infection

A second problem addressed is the need of high quality spatial-temporal data of hospital infection spread. To this end, we have proposed and implemented a simulation model and its simulator. The objective of the simulation model is to generate clinically realistic spatial-temporal data of infection outbreaks within hospitals. We adopted an agent-oriented approach combining spatial-temporal logic of the agents together with clinical semantic obtained from external epidemiological models. The main objectives of the simulation are the movements of patients through a hospital, and the infection and transmission of an MDR-bacteria disease. For this reason, our model consists of one type of agent: patients. To study the evolution of patient's infection, each one is associated with a health state. This state is an abstraction of their situation at any given time so as to adapt it to the SEIRD epidemic model (Fig. 1).

Agents can interact with each other if they share a room, and they can also interact with the environment. A contagion can occur in these interactions if one of the patients is infected or if the environment is contaminated.

For our hospital model, we have considered a two-story hospital, where we take into account the most likely areas in which a hospitalized patient can become infected: emergency room, radiology rooms, operating rooms, ward rooms and the Intensive Care Unit.

The simulations follow a discrete-time step-based approach. In each step of the simulation, patients can move from their current location to an available place in the hospital. Our proposal also simulates admissions and discharges. In addition to this, the health status of patients can change and the different places within the hospital can be contaminated or disinfected.

Our preliminary experiments have been conducted considering Clostridium Difficile infection (CDI) since it is the main cause of infectious diarrhea in hospitalized patients. Both its incidence and the severity of clinical manifestations have increased notoriously in recent years [9]. In order to obtain a fair comparative with state-of-the-art proposals, we have based part of our methodology on [5,6,12]. The results from these experiments are outlined in Fig. 2.

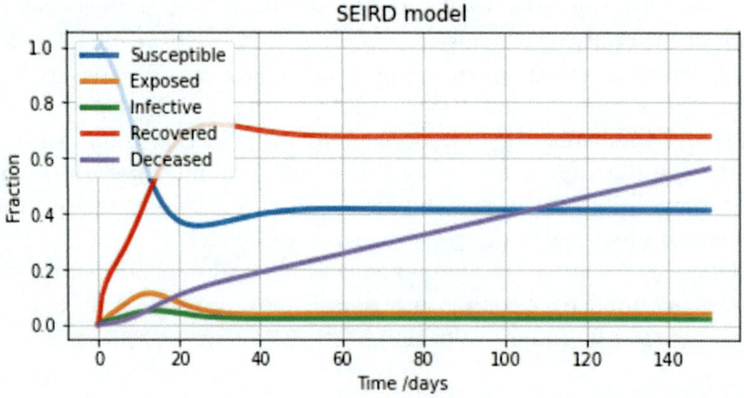

Fig. 1. SEIRD epidemic model.

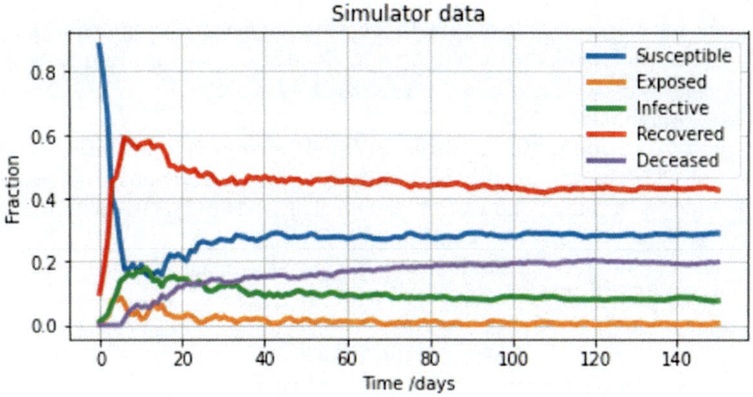

Fig. 2. Simulator results.

4 Current State

This thesis project started in October 2020 and we are reaching the halfway point of work. The current pandemic scenario has slowed down certain activities, particularly advice from the medical team.

We are waiting to publish the systematic review and we expect to have results from the simulator in the coming weeks. We are starting to run some tests to visualize the data output that we are getting from the simulator.

Acknowledgements. This work was partially funded by the SITSUS project (Ref: RTI2018-094832-B-I00), the CONFAINCE project (Ref: PID2021-122194OB-I00), supported by the Spanish Ministry of Science and Innovation, the Spanish Agency for Research (MCIN/AEI/10.13039/501100011033) and, as appropriate, by ERDF A way of making Europe. This research is also partially funded by the FPI program grant (Ref:PRE2019-089806).

References

1. Abat, C., Chaudet, H., Rolain, J.M., Colson, P., Raoult, D.: Traditional and syndromic surveillance of infectious diseases and pathogens. Int. J. Infect. Dis.: IJID: Official Publ. Int. Soc. Infect. Dis. **48**, 22–28 (2016). https://doi.org/10.1016/j.ijid.2016.04.021
2. Arantes, A., Carvalho, E.d.S., Medeiros, E.A.S., Farhat, C.K., Mantese, O.C.: Use of statistical process control charts in the epidemiological surveillance of nosocomial infections. Revista De Saude Publica **37**(6), 768–774 (2003). https://doi.org/10.1590/s0034-89102003000600012
3. Baumgartl, T., Petzold, M., Wunderlich, M., Höhn, M., Archambault, D., Lieser, M., Dalpke, A., Scheithauer, S., Marschollek, M., Eichel, V.M., Mutters, N.T., Consortium, H., von Landesberger, T.: In search of patient zero: visual analytics of pathogen transmission pathways in hospitals. IEEE Trans. Vis. Comput. Graph. **27**(2), 711–721 (2021). https://doi.org/10.1109/TVCG.2020.3030437,http://arxiv.org/abs/2008.09552, arXiv: 2008.09552

4. Centers for Disease Control and Prevention: General Recommendations for Routine Prevention and Control of mdros in Healthcare Settings (2019). https://www.cdc.gov/infectioncontrol/guidelines/mdro/table3-1-routine-prevention.html. Last access 9 May 2022
5. Clabots, C.R., Johnson, S., Olson, M.M., Peterson, L.R., Gerding, D.N.: Acquisition of Clostridium difficile by hospitalized patients: evidence for colonized new admissions as a source of infection. J. Infect. Dis. **166**(3), 561–567 (1992). https://doi.org/10.1093/infdis/166.3.561
6. Codella, J., Safdar, N., Heffernan, R., Alagoz, O.: An agent-based simulation model for Clostridium difficile infection control. Med. Decis. Making: Int. J. Soc. Med. Decis. Making **35**(2), 211–229 (2015). https://doi.org/10.1177/0272989X14545788
7. van Duin, D., Paterson, D.L.: Multidrug-resistant bacteria in the community. Infect. Dis. Clin. North Am. **30**(2), 377–390 (2016). https://doi.org/10.1016/j.idc.2016.02.004
8. Kim, D., Campos, M., Juarez, J.M., Canovas-Segura, B.: Visualization of spatial-temporal epidemiological data: a systematic review. J. Med. Internet Res. (2022). (under review)
9. Lital Meyer, S., Ricardo Espinoza, A., Rodrigo Quera, P.: Infección por clostridium difficile: epidemiología, diagnóstico y estrategias terapéuticas. Revista Médica Clínica Las Condes **25**(3), 473–484 (2014)
10. Moher, D., Liberati, A., Tetzlaff, J., Altman, D.G.: PRISMA Group: preferred reporting items for systematic reviews and meta-analyses: the PRISMA statement. PLoS Med. **6**(7), e1000097 (2009). https://doi.org/10.1371/journal.pmed.1000097
11. Myall, A.C., Peach, R.L., Weiße, A.Y., Davies, F., Mookerjee, S., Holmes, A., Barahona, M.: Network Memory in the Movement of Hospital Patients Carrying Drug-Resistant Bacteria. arXiv preprint arXiv:2009.14480 (2020)
12. Olson, M.M., Shanholtzer, C.J., Lee, J.T., Gerding, D.N.: Ten years of prospective Clostridium difficile-associated disease surveillance and treatment at the Minneapolis VA Medical Center, 1982–1991. Infection Control Hosp. Epidemiol. **15**(6), 371–381 (1994). https://doi.org/10.1086/646934
13. Tseng, Y.J., Wu, J.H., Ping, X.O., Lin, H.C., Chen, Y.Y., Shang, R.J., Chen, M.Y., Lai, F., Chen, Y.C.: A web-based multidrug-resistant organisms surveillance and outbreak detection system with rule-based classification and clustering. J. Med. Internet Res. **14**(5), e131 (2012). https://doi.org/10.2196/jmir.2056
14. World Health Organization: Antimicrobial Resistance (2021). https://www.who.int/news-room/fact-sheets/detail/antimicrobial-resistance. Last access 9 May 2022

An Intelligent and Green E-healthcare Model for an Early Diagnosis of Medical Images as an IoMT Application

Ibrahim Dhaini(✉), Soha Rawas, and Ali El-Zaart

Department of Mathematics and Computer Science, Faculty of Science, Beirut Arab University, Beirut, Lebanon
ibrahim.dhaini.bau@gmail.com, {soha.rawas2,elzaart}@bau.edu.lb

Abstract. The Internet of Things (IoT) is a fast-evolving technology that utilizes software, hardware, and computer devices to form a network of interconnected gadgets. IoMT integrates medical equipment and applications linked to health-care IT systems in an IoT-based ecosystem. Moreover, IoMT for health care is a massive data generator produced by sensors or any medical device attached to the Internet. As a result, transferring IoMT data to remote cloud databases is a popular procedure. This research proposes an intelligent and green e-healthcare model for an early diagnosis of medical images. Moreover, the research focuses on image segmentation, an essential phase in image analysis, and presents a precise and robust segmentation model. Furthermore, the research considers the high energy consumption of transferring massive data through the cloud. It suggests a new energy-aware VM placement model in a fog-based environment.

Keywords: E-healthcare system · IoMT · Fog computing · Image segmentation · Green computing

1 Introduction

The Internet of things is a technology that is infiltrating every part of our lives, including economics, smart cities, and healthcare, to name a few. The Internet of Things (IoT) is a group of heterogeneous network devices that may be linked together to generate and share data with other devices or people [1]. Internet of Medical Things (IoMT) refers to integrating medical equipment and applications linked to healthcare IT systems in an IoT-based ecosystem [2]. IoMT plays a vital role in providing patients with medical services in real-time regardless of their residence, age, or financial status. The Internet of medical things is a massive data generator produced by sensors, surveillance cameras, or any medical device attached to the Internet. Cloud infrastructure can overcome the storage and processing limitations of the IoT and provide the capabilities to store and process vast amounts of data. However, a real problem related to data latency transmission will arise in smart devices and applications that rely on real-time computation. To overcome this issue, many authors integrate fog computing into this infrastructure.

J. M. Machado et al. (Eds.): DCAI 2022, LNNS 585, pp. 159–164, 2023.
https://doi.org/10.1007/978-3-031-23210-7_16

The overall goal of this paper is to propose a segmentation model and an energy-aware VM placement methodology for an Intelligent and Green E-healthcare Model for an Early Diagnosis of Medical Images as an IoMT. The proposed models are essential parts of an intelligent green e-healthcare model for an early diagnosis of Medical images as an IoMT application. The e-healthcare model integrates the advantages of three technologies: the Internet of Things, Fog computing, and Cloud computing. The considered framework consists of the Image Acquisition, Fog, and Cloud layers. The following are the significant commitments:

1. Develop a precise and robust segmentation model.
2. Develop and implement a novel energy-aware VM placement model that considers Energy cost + communication cost + VM cost to boost the QoE.

2 Motivation

Most developing countries suffer from a shortage of doctors and medical staff. Citizens in these countries are affected by different crises, depriving them of basic life needs, including healthcare, food, electricity, and other necessities. Taking advantage of technological advancements, particularly in the Internet of Things, encourages researchers to investigate the prospect of using these technologies to alleviate citizen burdens. Implementing an intelligent e-healthcare model for an early diagnosis of Medical images as an IoMT application can support people. This e-healthcare system needs to be green which means using less energy than is typical. Moreover, any suggested medical system must afford real-time responses to contribute to saving people's lives thoughtfully. Cloud network latency may be a cause to prevent achieving this goal. These issues necessitate that, in addition to being intelligent, the proposed system must include green-aware network technologies.

3 Objectives and Aim of Study

The main objectives of this paper are:

- Developing a precise and robust segmentation model
- Developing and implementing a novel green VM placement model

These two objectives will help in the future:

- To create a healthcare system that may be accessed anytime, from any location, and by anyone.
- To early diagnose medical images and detect any anomalies.

4 Related Work

This paper examines a group of publications for authors in the healthcare field using IoT. Breast cancer was investigated by Suresh et al. [3] through an automatic real-time classifier system based on IoT infrastructure. Image processing methods were applied to segment images and extract features from them. Khan et al. [4] presented a method for classifying brain MR images into malignant and non-cancerous situations. The method combines with IoT device technology to detect aberrant brain cells early and aid doctors in their diagnoses. Palani et al. propounded a system for predicting lung cancer from (CT) medical images submitted through an internet application that works with IoT devices. Along with the application, a new segmentation algorithm was proposed [5]. The author did not mention how the proposed system was fully automated or implemented and how it could collect the images from users. In their paper, Kaur et al. [6] tested various machine learning approaches on different datasets to predict groups of diseases. Their work suggested a remote monitoring system for detecting diseases using random forest classifiers and IoT. The system sends the collected medical data to the cloud for storage and automatically analyzes images. The doctors can observe the collected medical data at any time and write their observations in a cloud database. The authors focus on the machine learning approaches in this work rather than implementing the other phases of the system.

The integrated phases introduced in this proposal make the suggested intelligent e-healthcare system different from those mentioned above. None of the reviewed systems authors have implemented all of the system capabilities described in their publications. Some authors use cloud servers as storage and classification processes, but the energy consumption is not within their work scope. Others suggested an integrated e-healthcare system with the help of fog and cloud technology but not for medical images. This proposal will implement a green intelligent e-healthcare model for the early detection of medical images as an IoMT application.

5 E-healthcare Model

The image segmentation and VM placement models proposed in the latter two sections are essential parts of an intelligent and green e-healthcare model for an early diagnosis of Medical images. The introduced system, see Fig. 1, comprises the image acquisitions layer, fog layer, and cloud layer.

5.1 Image Acquisition Layer

In this part of the system, medical images are gathered from various sources and sent to the fog layer for analysis and classification. This layer consists of two components: registration and medical image uploading. The first registration step is to sign up for the system via a mobile application or website. Healthcare centers or doctors can register patients to the system. The software keeps patients' information in a secure database once their information enters the system. The second component is the medical image uploading. A healthcare architecture that deals with imaging can process different sources of images. Medical personnel who are part of this health system should upload medical photos using smartphones or computers.

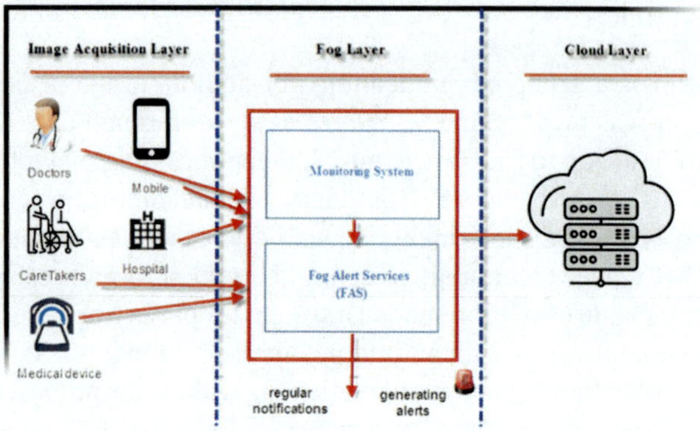

Fig. 1. Suggested E-healthcare system

5.2 Fog Layer

The real benefit of incorporating fog computing into an IoMT system is that it allows computer resources like data management, networking, processing, and storage to be closer to the Things. Locating resources near IoT devices mitigate the network latency in cloud computing. Furthermore, fog computing can afford real-time applications and speedy responses rather than requiring a lengthy trip around the cloud. This layer comprises two parts: Monitoring System and Fog Alert Service (FAS). The monitoring System part is responsible for image preprocessing, segmenting, and later classifying it as diseased or safe. Moreover, the patients' records and extracted features have to go directly to cloud servers so the doctor can diagnose them. The Fog Alert Service (FAS), based on the classification result in the first component, can issue an alert to the physician.

5.3 Cloud Layer

Patients' records, medical images, image diagnostic, alerts, and notifications are all saved in the cloud. Doctors can investigate a specific medical condition anytime and from any location using this process. Patients with data saved in the cloud can conveniently consult with other system-approved doctors.

6 Proposed Image Segmentation Model

In order to use medical images in disease diagnosis, different aspects of image analysis must be completed: Image Acquisition, Image Enhancement, Image Segmentation, Feature Extraction, Image Classification, and finally, image diagnosis. Image segmentation is a fundamental step for many image analysis and preprocessing tasks. Image segmentation is the technique of partitioning an image into different groups with comparable features, such as shape, color and gray percentage, and texture [7]. Image segmentation methods are divided into four categories: region- and boundary-based, thresholding, and hybrid approaches [8]. Thresholding approaches garnered a lot of attention and were frequently employed by researchers because of their simplicity, ease of customization, and high processing speed. In 1993 [9], Li and Lee proposed a sequential approach to find the optimal threshold value using MCET based on Gaussian distribution. Li et al. suggested a method to obtain the optimal threshold rate (t*) by utilizing minimum cross entropy thresholding using Gaussian distribution. The MCET of Li estimates the optimal thresholding value t* by minimizing the cross-entropy.

$$t^* = \arg\min_t(D(I, I_t)) = \arg\min_t(D(t)) \tag{1}$$

The cross-entropy is calculated between the initial image I(x, y) and the threshold image $I_t(x, y)$ using the following formula:

$$D(I, I_t) = \sum_{i=0}^{t} i.h(i).\log\frac{i}{\mu_A} + \sum_{i=t+1}^{L} i.h(i).\log\frac{i}{\mu_B} \tag{2}$$

where t is the obtained threshold, h(i) is the image histogram value, $\mu_a(t)$ and $\mu_b(t)$ are the mean values of the object and background classes in the image.

In [8] paper, Rawas et al. proposed a novel segmentation model to improve Li and Lee's work. Their approach was based on the minimum cross-entropy thresholding method that Li et al. introduced using combinations of heterogeneous distribution. In 2022 [10], Rawas and El-Zaart proposed a novel segmentation model that used a derivation of a hybrid cross entropy thresholding technique. Our first contribution is to design and implement a proposed segmentation model to improve Rawas and El-Zaart's work based on the MCET using heterogeneous distributions.

7 Proposed Green VM Placement Model

According to studies, servers with low computational pressure waste 70% of their energy consumption. In addition, 80 percent of working nodes in data centers use less than half of their CPU capacity [11]. Server consolidation, which minimizes the number of active running servers, saves energy in data centers. Cloud computing utilizes virtualization technology and divides the physical resources of one or more computer nodes into numerous execution applications known as Virtual Machines [12]. VM migration from one physical node to another necessitates the use of a method to choose which physical node to use; this process is known as VM placement [12]. VM placement plays a vital role in server consolidation by grouping specific virtual machines on the least number of physical machines while putting the other physical machines into sleep mode. Rawas [11] proposed a novel algorithm (ENAV) to handle the VM placement problem in a cloud-based environment. Unlike other suggested methods to the VM placement problem, the ENAV model considers the energy consumption of the physical servers, the cost of communicating and transferring data between VMs, and the cost of the VMs.

The use of fog computing in this suggested model reduces the amount of data that crosses the network, lowering energy costs. However, applying fog architecture is not the only suggestion for lowering energy use. A new energy-aware VM placement model in a fog-based environment will be developed and implemented while considering the physical servers' energy consumption, the cost of communicating and transferring data between VMs, and the cost of the VMs. The proposed approach aims to perform the Virtual Machine Placement model so that it consumes the least amount of energy while also meeting customers' Service Level Agreement (SLA) and Quality of Experience (QoE) requirements.

References

1. Andriopoulou, F., Dagiuklas, T., Orphanoudakis, T.: Integrating IoT and fog computing for healthcare service delivery. In: Components and Services for IoT Platforms, pp. 213–232. Springer, Heidelberg (2017)
2. Kashani, M.H., et al.: A systematic review of IoT in healthcare: applications, techniques, and trends. J. Netw. Comput. Appl., 103164 (2021)
3. Suresh, A., Udendhran, R., Balamurgan, M., Varatharajan, R.: A novel Internet of Things framework integrated with real time monitoring for intelligent healthcare environment. J. Med. Syst. **43**(6), 1 (2019). https://doi.org/10.1007/s10916-019-1302-9
4. Khan, S.R., et al.: IoMT-based computational approach for detecting brain tumor. Futur. Gener. Comput. Syst. **109**, 360–367 (2020)
5. Palani, D., Venkatalakshmi, K.: An IoT based predictive modelling for predicting lung cancer using fuzzy cluster based segmentation and classification. J. Med. Syst. **43**(2), 1–12 (2019)
6. Kaur, P., Kumar, R., Kumar, M.: A healthcare monitoring system using random forest and internet of things (IoT). Multimedia Tools Appl. **78**(14), 19905–19916 (2019). https://doi.org/10.1007/s11042-019-7327-8
7. Wang, E.K., et al.: A deep learning based medical image segmentation technique in Internet-of-Medical-Things domain. Futur. Gener. Comput. Syst. **108**, 135–144 (2020)
8. Rawas, S., El-Zaart, A.: Precise and parallel segmentation model (PPSM) via MCET using hybrid distributions. In: Applied Computing and Informatics (2020)
9. Li, C.H., Lee, C.: Minimum cross entropy thresholding. Pattern Recogn. **26**(4), 617–625 (1993)
10. Rawas, S., El-Zaart, A.: Towards an Early Diagnosis of Alzheimer Disease: A Precise and Parallel Image Segmentation Approach Via Derived Hybrid Cross Entropy Thresholding Method (2022)
11. Rawas, S.: Energy, network, and application-aware virtual machine placement model in SDN-enabled large scale cloud data centers. Multimedia Tools Appl. **80**(10), 15541–15562 (2021). https://doi.org/10.1007/s11042-021-10616-6
12. Masdari, M., Nabavi, S.S., Ahmadi, V.: An overview of virtual machine placement schemes in cloud computing. J. Netw. Comput. Appl. **66**, 106–127 (2016)

Towards Highly Performant Context Awareness in the Internet of Things

Elias Werner(✉)

Technische Universität Dresden, Center for Information Services and High Performance Computing (ZIH), 01062, Dresden, Germany
elias.werner@tu-dresden.de

Abstract. Context awareness and adaption are an important requirement for data streaming applications in the Internet of Things (IoT). Concept drift arises under changing context and many solutions for drift detection and system adaption have been developed in the past. Even though the computing resources of these approaches are not negligible, performance aspects, e.g. runtime or memory usage and scalability have not been investigated adequately. The goal of this thesis is to fill this gap and to provide a mean for performance investigations of concept drift handling methods and other approaches that are developed in the research field Data Science. Therefore performance benchmarking of state-of-the-art solutions is conducted. Moreover the thesis discusses performance bottlenecks of the approaches and demonstrates possible improvements based on the knowledge gained by leveraging tools and methods from the performance analysis domain. Approaches are then implemented into an IoT application and deployed with stream processing engines for a final evaluation.

Keywords: Concept drift detection · Performance analysis · IoT

1 Introduction

In the last years, the amount of collected data increased significantly and is expected to reach about 175 ZB in the year 2025. This data is gathered by many versatile source devices that operate under different conditions and contextual settings, forming the Internet of Things (IoT). The contexts of the data sources affect the collected data and thus is implicitly represented in the data values. However, system may assume the same context for data or do not consider the context of the data accordingly and thus lead to wrong decisions, e.g. model predictions. The contextual differences of the data values and the related change of the data distribution is considered as concept drift. Therefore it is crucial for future reliable artificial intelligence (AI) and IoT systems to consider concept drift. Many approaches to deal with that problem have been developed in the past. However, these approaches focus on data science (DS) related *quality measures*, e.g. accuracy, precision, recall. To the best of the author's knowledge, non

J. M. Machado et al. (Eds.): DCAI 2022, LNNS 585, pp. 165–170, 2023.
https://doi.org/10.1007/978-3-031-23210-7_17

of the related studies focuses on *performance measures*, e.g. runtime, memory usage or the distribution across IoT actors of the developed approaches. Nevertheless, this *performance* related considerations are important for IoT scenarios such that future AI systems are applicable on the upcoming big data requirements, e.g. real-time responsiveness or energy efficiency.

Therefore, the contribution of the thesis aims to fill this gap by investigating state-of-the-art approaches for concept drift handling from a *performance* point of view, highlighting bottlenecks of related approaches and presenting solutions for scaling and distributing mechanisms for concept drift handling in IoT infrastructure. The present paper gives an outline for the thesis and is structured as follows: in Sect. 2, selected state-of-the-art solutions for concept drift detection are presented. Section 3 presents an overview of existing approaches to conduct performance investigations in DS. Section 4 proposes the pursued research approach and Sect. 5 concludes the paper.

2 Existing Solutions for Concept Drift Detection

Many different approaches have been developed in the last years to detect concept drift. On the one hand, there are supervised solutions available such as DDM [5] or Adwin [2]. This solutions rely on the presence of labeled data and therefore are only partially applicable in an IoT setting, since data labels might not be available. On the other hand, unsupervised methods, such as presented by Haque et al. [9], Mustafa et al. [15] or Kim et al. [12] have been developed that work either on the gathered raw data or consider the output data of a pre-trained machine learning (ML) model to detect a concept drift.

For both, supervised and unsupervised approaches, the evaluation in the papers in terms of quality (e.g. accuracy or recall) is extensive and detailed. Moreover there exist survey papers, that conduct comprehensive evaluation of the approaches. On the performance side (e.g. runtime or memory usage), there are only a few works that consider the runtime of their approaches (dos Reis et al. [19], Pinagé et al. [16]). For the supervised case, there exist a few papers that focus on the parallelization of the implementations, e.g. for Adwin [8]. Nevertheless, comprehensive studies on the performance aspects or the scalability of the developed solutions are not considered accordingly. Moreover, to the best of the author's knowledge, there exists no solution for the unsupervised case of concept drift detection, i.e. detecting concept drift without knowing the true label of the data. The lack on research in this field is emphasized by the fact that there is no survey paper comparing concept drift detection techniques (DDT) from a performance point of view or discussing the scalability. Therefore the author has to conclude that it is required to investigate the performance (e.g. runtime or memory usage) of state-of-the-art DDTs to determine if the available solutions are ready for IoT settings with versatile data sources, large datasets and real-time requirements. Moreover, recent surveys affirm the importance of such performance considerations for concept drift handling approaches and emphasize the relevance of this aspects for DS methods in general [6]. The need for performance investigations and optimisation is further emphasised by the sheer existence of benchmarks such as MLPerf [14]. Unfortunately there is a lack of knowledge and accessibility of prevalent Performance analysis (PA) tools in the DS domain.

3 Performance Investigations in Data Science

Performance analysis (PA) is a mean to consider performance aspects such as runtime or memory usage of an application and to get insights into a programs behavior. This supports developers in analyzing and finding bottlenecks in the program and can help to improve the application. In this section we describe preceding developments in the field of PA with respect to DS. We highlight that the DS community lacks in support of tools to conduct performance investigations and present an early approach that bridges between the two domains.

The popular editor for DS applications Jupyter comes with some internal commands for rudimentary performance investigations. So-called magic commands [10], e. g. `%prun` or `%timeit`, provide runtimes or number of function calls, but do not provide the functions' context, such as thread or the calling function. Furthermore, the `%prun` command does not provide a graphical representation and relies on the Python profiler [18], so it is not usable for non-Python functions, such as CUDA. Few of the popular frameworks for DS provide means to investigate performance of applications, e. g. the TensorFlow Profiler [20]. However, the features they provide, are specific for that particular framework. Therefore, performance tools cannot be reused, which results in different aspects shown to a performance analyst, when working with a different framework.

A solution for collecting performance data in a standardised format would allow to build or reuse tools. The Score-P framework [13] provides performance data in a format that can be used by several different analysis or visualisation tools in order to be able to use the specific strengths of each particular tool. Due to its development with a focus on HPC applications, Score-P is scalable and produces performance data in several formats in order to support further analysis via tools such as Vampir or Cube among others. Moreover, Score-P supports not only C, C++ and Fortran, but also binds to Java [4] and Python [7] applications, making it a good candidate for holistic performance analysis.

As one approach to bridge between the two domains PA and DS, traditional PA tools have been integrated in Jupyter. Therefore we developed a custom Jupyter kernel that links to Score-P [21] by using the Score-P Python bindings [7] and the Jupyter wrapper kernel interface [11]. When a user executes code in Jupyter and uses the developed kernel, performance data will be emitted for further investigations of the runtime or memory usage of code segments.

4 Proposing a Research Approach

Adapting to concept drift is a major requirement for IoT systems. In Sect. 2 it was highlighted, that state-of-the-art DDTs need to be investigated from a performance point of view since there is no evidence for sufficient performance or scalability of the approaches for future IoT challenges. Means to focus on performance aspects of an application are provided by PA techniques and tools. Even though the developed Score-P kernel for Jupyter as described in Sect. 3 is a first step towards bridging between the PA and DS domain, further investigations to make DS applications ready for IoT challenges are required. The aim of this thesis is to fill this gap for the concept drift requirement. Hence, the contributions are manifold. A general overview of the several research components is depicted in Fig. 1 and explained next.

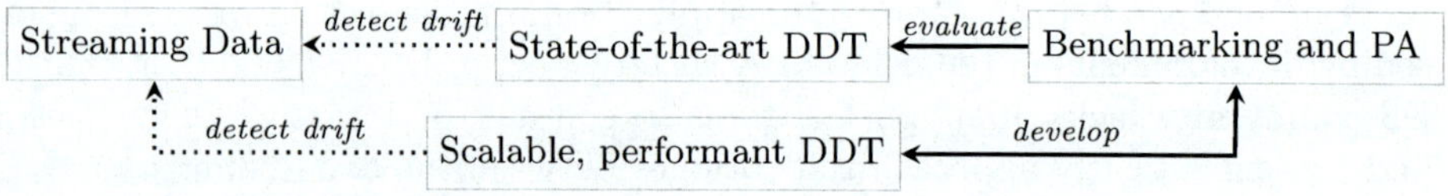

Fig. 1. Brief overview of research approach. State-of-the-art DDTs are evaluated and based on the gained knowledge, scalable and performant solutions are developed.

Benchmarking state-of-the-art approaches for concept drift detection: In order to evaluate and compare DDTs that operate on streaming data in the IoT, a benchmark suite is required to conduct comprehensive evaluation of the approaches from a performance point of view. A challenge is, to combine the performance aspects with the quality aspects, e.g. which concept drift detector achieves the highest accuracy within a certain time or with limited memory? Therefore a benchmark based on the MOA framework [3] is developed to create comparative data streaming settings for the investigations. MOA is a Java based application for creating generic data streams that allows the evaluation of AI algorithms in this domain. It also supports the introduction of different concepts but needs to be adapted for the proposed performance investigations.
Providing a mean for PA in DS: As outlined in Sect. 3, one contribution is the developed Jupyter kernel that binds to Score-P and enables performance investigations of the developed algorithms. Since the publication of the kernel [21], support for multi-cell PA was added. In the future, the kernel will be improved by reducing the introduced overhead and supporting visualizations of the gathered performance data directly in the Jupyter interface.
Proposing solutions for scalable concept drift detectors: Based on the benchmark results and the investigations with the Score-P Jupyter kernel, it is possible to get insights into the runtime behaviour of the considered concept drift handling approaches and to develop solutions for parallel or distributed computing. Therefore also approaches to deploy algorithms via the map-reduce paradigm need to be investigated.

Furthermore, investigations of the following research directions will be considered:
Visualizing concept drift and DDTs: The visualization of concept drift and the underlying data is one step towards understanding AI systems and the gathered

data. However, to the best of the author's knowledge, there are no visualizations available focusing on concept drift. Early approaches such as presented by Pratt et al. [17] need to be adapted to today's big data requirements but can support future developments.

Automated algorithm selection for concept drift detection: Barros et al. [1] evaluated supervised concept drift detectors from a quality point of view. They conclude that the approaches $RDMA$ and $HDDM_a$ achieve the best quality on average in multiple experimental settings. However, some scenarios of their experiments also highlighted other approaches in terms of high quality. Therefore it is of interest, whether an ensemble of different solutions for concept drift handling achieves even better results. Such an approach could then also apply automated algorithm selection to provide the best concept drift detector for a certain setting.

5 Conclusion and Outlook

In this work the need for performance investigations of algorithms for concept drift handling was motivated. Therefore state-of-the-art approaches for concept drift handling have been introduced and the lack of performance considerations has been highlighted. Moreover existing solutions for performance investigations in DS have been discussed. Based on this considerations, the paper highlights the pursued research directions that aim to fill the presented research gaps.

While the Score-P Python Kernel for Jupyter is already published [21], experiments based on the kernel to evaluate state-of-the-art concept drift detectors are conducted. At next, it is planned to provide the achieved findings to the public and publish a benchmark infrastructure for comparison of future solutions.

Acknowledgment. This work was supported by the German Federal Ministry of Education and Research (BMBF, 01/S18026A-F) by funding the competence center for Big Data and AI "ScaDS.AI Dresden/Leipzig". The authors gratefully acknowledge the GWK support for funding this project by providing computing time through the Center for Information Services and HPC (ZIH) at TU Dresden.

References

1. Barros, R.S.M., Santos, S.G.T.C.: A large-scale comparison of concept drift detectors. Inf. Sci. **451**, 348–370 (2018)
2. Bifet, A., Gavalda, R.: Learning from time-changing data with adaptive windowing. In: Proceedings of the 2007 SIAM International Conference on Data Mining, pp. 443–448. SIAM (2007)
3. Bifet, A., Holmes, G., Pfahringer, B., Kranen, P., Kremer, H., Jansen, T., Seidl, T.: Moa: Massive online analysis, a framework for stream classification and clustering. In: Proceedings of the First Workshop on Applications of Pattern Analysis, pp. 44–50, PMLR (2010)

4. Frenzel, J., Feldhoff, K., Jaekel, R., Mueller-Pfefferkorn, R.: Tracing of multithreaded java applications in score-p using bytecode instrumentation. In: ARCS Workshop 2018; 31th International Conference on Architecture of Computing Systems, pp. 1–8, VDE (2018)
5. Gama, J., Medas, P., Castillo, G., Rodrigues, P.: Learning with drift detection. In: Brazilian Symposium on Artificial Intelligence, pp. 286–295. Springer, Heidelberg (2004)
6. Gemaque, R.N., Costa, A.F.J., Giusti, R., Dos Santos, E.M.: An overview of unsupervised drift detection methods. Wiley Interdisc. Rev.: Data Mining Knowl. Discov. **10**(6), e1381 (2020)
7. Gocht, A., Schöne, R., Frenzel, J.: Advanced python performance monitoring with score-p. In: Tools for High Performance Computing 2018/2019, pp. 261–270. Springer, Heidelberg (2021)
8. Grulich, P.M., Saitenmacher, R., Traub, J., Breß, S., Rabl, T., Markl, V.: Scalable detection of concept drifts on data streams with parallel adaptive windowing. In: EDBT, pp. 477–480 (2018)
9. Haque, A., Khan, L., Baron, M.: Sand: Semi-supervised adaptive novel class detection and classification over data stream. In: Proceedings of the Thirtieth AAAI Conference on Artificial Intelligence, p. 1652–1658, AAAI'16. AAAI Press (2016)
10. Ipython Built-in Magic Commands. https://ipython.readthedocs.io/en/stable/interactive/magics.html
11. Jupyter Documentation. https://jupyter.readthedocs.io/en/latest/projects/architecture/content-architecture.html
12. Kim, Y., Park, C.H.: An efficient concept drift detection method for streaming data under limited labeling. IEICE Trans. Inf. Syst. **100**(10), 2537–2546 (2017)
13. Knüpfer, A., Rössel, C., Biersdorff, S., Diethelm, K., Eschweiler, D., Geimer, M., Gerndt, M., Lorenz, D., Malony, A., Nagel, W.E., et al.: Score-p: a joint performance measurement run-time infrastructure for periscope, scalasca, tau, and vampir. In: Tools for High Performance Computing 2011, pp. 79–91. Springer, Heidelberg (2012)
14. Mattson, P., Cheng, C., Diamos, G., Coleman, C., Micikevicius, P., Patterson, D., Tang, H., Wei, G.Y., Bailis, P., Bittorf, V., et al.: Mlperf training benchmark. Proc. Mach. Learn. Syst. **2**, 336–349 (2020)
15. Mustafa, A.M., Ayoade, G., Al-Naami, K., Khan, L., Hamlen, K.W., Thuraisingham, B., Araujo, F.: Unsupervised deep embedding for novel class detection over data stream. In: 2017 IEEE International Conference on Big Data (Big Data), pp. 1830–1839. IEEE (2017)
16. Pinagé, F., dos Santos, E.M., Gama, J.: A drift detection method based on dynamic classifier selection. Data Mining Knowl. Discov. **34**(1), 50–74 (2020)
17. Pratt, K.B., Tschapek, G.: Visualizing concept drift. In: Proceedings of the Ninth ACM SIGKDD International Conference on Knowledge Discovery and Data Mining, pp. 735–740 (2003)
18. Python Profiler Library. https://docs.python.org/3/library/profile.html
19. dos Reis, D.M., Flach, P., Matwin, S., Batista, G.: Fast unsupervised online drift detection using incremental Kolmogorov-Smirnov test. In: Proceedings of the 22nd ACM SIGKDD International Conference on Knowledge Discovery and Data Mining, pp. 1545–1554 (2016)
20. Tensorflow Profiler. https://www.tensorflow.org/guide/profiler
21. Werner, E., Manjunath, L., Frenzel, J., Torge, S.: Bridging between data science and performance analysis: tracing of jupyter notebooks. In: The First International Conference on AI-ML-Systems, pp. 1–7 (2021)

Adaptive System to Manage User Comfort Preferences and Conflicts at Everyday Environments

Pedro Filipe Oliveira[1,2(✉)], Paulo Novais[2], and Paulo Matos[1]

[1] Instituto Politécnico de Bragança, Campus de Santa Apolónia, 5300-253 Bragança, Portugal
pmatos@ipb.pt

[2] Department of Informatics, Algoritmi Centre/University of Minho, Braga, Portugal
poliveira@ipb.pt, pjon@di.uminho.pt

Abstract. Nowadays an actual problem on IoT adaptive systems is to manage user preferences and local actuators specifications. This paper uses a multi agent system to achieve a Adaptive Environment System, that supports interaction between persons and physical spaces, where spaces smartly adapt to their user preferences in a transparent way. This work has been developed using a multi agent system architecture with different features to achieve a solution that reach all the proposed objectives.

Keywords: Adaptive-system · AmI · Multi-agent · IoT · Actuators · Preferences · Constraints

1 Introduction

Artificial Intelligence field continues with an exponential growth rate, and multi-agent systems have been used to solve several situations, related to Ambient Intelligence.

Ambient Intelligence (AMI), is an ubiquitous, electronic and intelligent environment, recognized by different technologies/systems interconnection, in order to carry out different daily tasks in a transparent and autonomous way for the user [1].

Thus, multi-agent systems are made up of autonomous agents present in the environment and who have the ability to make decisions derived from interpreted stimulus and connection with other agents, to achieve common goals [8].

Currently there are different languages and platforms for this type of systems development, namely *3APL*, *Jack*, *Jade/Jadex*, *Jason*, among others.

This work propose an autonomous Smart Home model controlled by cognitive agents using *Jason* and *ARGO* and to manage physical devices, since *ARGO* agents allow communication with different controllers (Arduino, Raspberry).

For this, the work has a prototype of a house with six divisions, each with lighting, and a heating system.

J. M. Machado et al. (Eds.): DCAI 2022, LNNS 585, pp. 171–176, 2023.
https://doi.org/10.1007/978-3-031-23210-7_18

The main expected contribution of this work is the possibility of applying MAS to ubiquitous prototypes using the *Jason* framework and *ARGO* architecture applied to intelligent environments.

2 Multi-agent System

2.1 Assumptions

To optimize the predictions of the solution proposed, an architecture for a multi-agent system was defined. The roles that each agent should represent, as well as the negotiation process to be taken, and the different scenarios in which this negotiation should take place and the way it should be processed were specified.

For the project development, two phases are defined as follows:

- Hardware (local systems) installation;
- Multi-agent system development;

Firstly, the entire physical structure must be prepared, where the local devices (Raspberry) equipped with the network technologies previously identified, so that they can detect the users present in the space.

The comfort preferences of each user present in the environment is sent to the agent every time an *ARGO* agent performs his reasoning cycle (by calling the *getPercepts* method, which must exist in all controllers that need to send perceptions to agents).

Thus, the MAS must be programmed independently from the hardware, taking into account only the actions that must be performed to achieve the ideal comfort values for the space in question, and then these values are sent to the actuators. The connection to the actuators was not taken into account in this work, implying that it is automatic and without any constraint for the user.

A prototype was thus implemented in a house, taking into account all the architecture of the MAS and the comfort actuators present in it.

For this purpose, a *Raspberry* is used per division, in this case three on the ground floor (living room/kitchen, office, bedroom) and three on the first floor (one at each environment).

Regarding the actuators, these divisions have a hydraulic radiant floor heating system heated by a heat pump, and a home automation system that controls the luminosity intensity in the different rooms.

This work proposes an autonomous Smart Home model, controlled through cognitive agents, which get the final information to be applied by actuators.

For do that, a six divisions house was prototyped with different comfort features, namely temperature, luminosity, audio and video. The considered parameters for performance evaluation are as follows:

- Number of agents used;
- Agent speed reasoning;
- Information filtering;
- Environment perception time.

2.2 Multi-agent System Architecture

ARGO is a customized Jason agent architecture to enable the programming of robotic and ubiquitous agents using different prototyping platforms.

ARGO allows cognitive agents and a real environment (using controllers) intermediation through *Javino* middleware, which communicates with hardware (sensors and actuators). In addition, use of BDI on robotic platforms can generate bottlenecks in perceptions processing and, consequently, unwanted execution delays, this extension also has a perception filtering mechanism at run time [7].

A MAS using *Jason* and *ARGO* can be made up of traditional *Jason* and *ARGO* agents that work simultaneously. *Jason* agents can carry out plans and actions only at software level and communicate with other agents in the system (including *ARGO* agents).

On the other hand a *ARGO* agent, is a traditional agent with additional characteristics, such as, for instance, the ability to communicate with the physical environment, perceive and modify it, and also filter the perceived information.

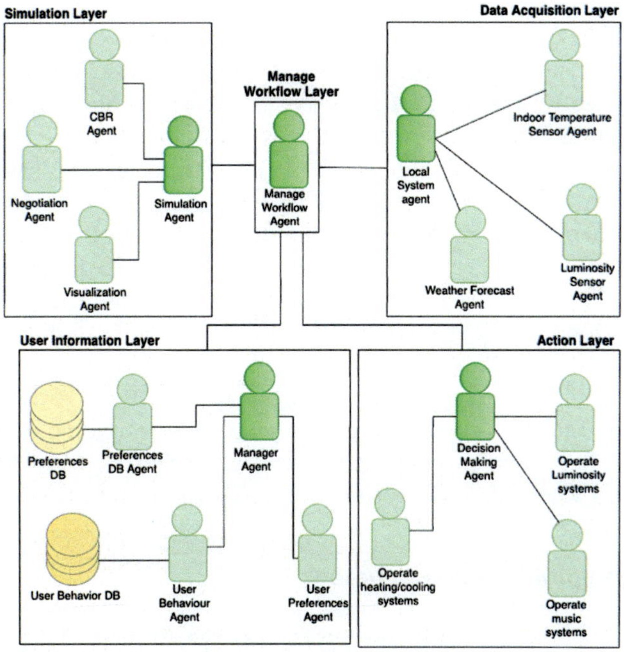

Fig. 1. Multi-agent system architecture

Figure 1 represent the different layers architecture separation, to easily identify the purpose of each, and agents containing it.

The layers description is as follows:

- **Data acquisition layer**, will import necessary information for the agents operation, namely information of interior and exterior temperature and light sensors.
- **User layer**, this layer has an agent that will represent each user and his preferences that must be used in the negotiation process.
- **Local System layer,** here each local system will be represented by an agent, which contains all necessary information to this location, either the referred to user preferences, or local/users security (maximum/minimum temperature, safety values for CO_2, etc.).
- **Simulation layer**, in this layer will be the negotiation between different agents involved, namely conflicts management between different users and local systems. After the negotiation process ends we will have as result the values to be applied at the environment.
- **Action layer**, after the simulation layer process execution, the values to be applied are obtained. These are used in this layer and sent to actuators that will apply them in the different automation systems and actuators present at the environment.

There will be one principal agent who will represent local system, namely each individual environment, where it was a need to ensure individualized comfort conditions, such as a room in a house, or a office in a building. This agent will take into account any directives that may exist for this environment, such as lower or upper limits to different comfort conditions, or also safety parameters that may be critical for a given space. This agent will have a obviously prevalence relative to others, since it will be the dominant for a given environment.

With users respect, each one in the space, will also be represented by an agent, this will receive user preferences from main system, for the place where it is, as well for the time in which it is. Also in this situation there will be a prioritization that identifies which user will have environment supremacy, so it also has an increase in the negotiation process.

In decision-making process, all users agents and agents representing the environment will be taken into account. With the different priorities that each of them has, and with this information will begin the negotiation process.

2.3 Multi-agent System Schema

The Multi-agent that supports the system was developed using JADE, and implements five different roles types to agents:

- Environment Agent: Provide information on the environment status. A new Environment Agent is created for each environment that is introduced into the system.
- Sensor Agent(s): Responsible for retrieve information of different conditions on the environment, namely temperature, brightness, and others depending of each environment conditions.
- Preference Agent: Keeps track of the preferences card.

- User Agent(s): Responsible for the negotiation process. Each user agent is associated with a single user present in the environment.
- Negotiation manager Agent: Created in each environment in order to manage the negotiation process between different User Agents.

At Fig. 2, is summarized the developed schema.

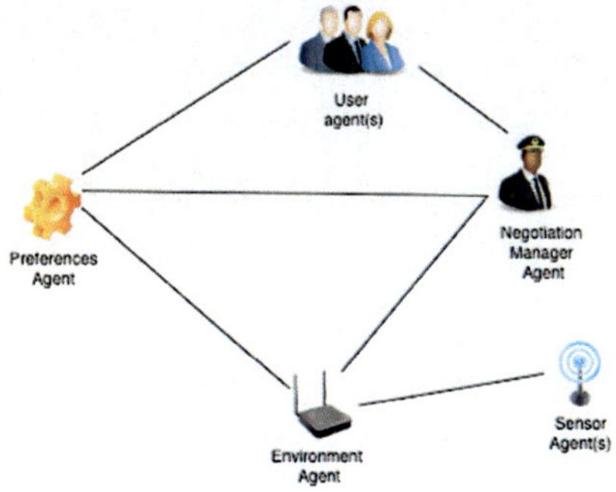

Fig. 2. Multi-agent system schema

3 Discussion and Conclusions

With this work, the specification of constraints for all specifications of proposed preferences for this work was achieved. In this way the safety of users and actuators present in space is achieved. The agent system modeling is fully developed. At this stage the agent layer is developed and implemented, and is now in a testing phase in the testing environment developed for this project.

For future work, the results of the testing phase, will be analyzed and evaluated and with that results improve this project and support other works in this field.

This work aims to give continuity and finalize the doctoral work presented in previous editions [2–6].

Acknowledgements. This work has been supported by FCT - Fundação para a Ciência e Tecnologia within the R&D Units Project Scope: UIDB/00319/2020.

References

1. Chaouche, A.C., Seghrouchni, A.E.F., Ilié, J.M., Saidouni, D.E.: A higher-order agent model with contextual planning management for ambient systems. In: Transactions on Computational Collective Intelligence XVI, pp. 146–169. Springer, Heidelberg (2014)

2. Oliveira, P., Matos, P., Novais, P.: Behaviour analysis in smart spaces. In: 2016 Intl IEEE Conferences on Ubiquitous Intelligence & Computing, Advanced and Trusted Computing, Scalable Computing and Communications, Cloud and Big Data Computing, Internet of People, and Smart World Congress (UIC/ATC/ScalCom/CBDCom/IoP/SmartWorld), pp. 880–887. IEEE (2016)
3. Oliveira, P., Novais, P., Matos, P.: Challenges in smart spaces: aware of users, preferences, behaviours and habits. In: International Conference on Practical Applications of Agents and Multi-Agent Systems, pp. 268–271. Springer, Heidelberg (2017)
4. Oliveira, P., Pedrosa, T., Novais, P., Matos, P.: Towards to secure an IoT adaptive environment system. In: International Symposium on Distributed Computing and Artificial Intelligence, pp. 349–352. Springer, Heidelberg (2018)
5. Oliveira, P.F., Novais, P., Matos, P.: A multi-agent system to manage users and spaces in a adaptive environment system. In: International Conference on Practical Applications of Agents and Multi-Agent Systems, pp. 330–333. Springer, Heidelberg (2019)
6. Oliveira, P.F., Novais, P., Matos, P.: Using Jason framework to develop a multi-agent system to manage users and spaces in an adaptive environment system. In: International Symposium on Ambient Intelligence, pp. 137–145. Springer, Heidelberg (2020)
7. Stabile, M.F., Sichman, J.S.: Evaluating perception filters in BDI Jason agents. In: 2015 Brazilian Conference on Intelligent Systems (BRACIS), pp. 116–121. IEEE (2015)
8. Wooldridge, M.: An Introduction to Multiagent Systems. Wiley, New York (2009)

Check for updates

ML-Based Automation of Constraint Satisfaction Model Transformation and Solver Configuration

Ilja Becker(✉), Sven Löffler, and Petra Hofstedt

Programming Languages and Compiler Construction Group,
Brandenburg University of Technology, 03046 Cottbus, Germany
{ilja.becker,sven.loeffler,hofstedt}@b-tu.de

Abstract. Constraint Programming is a powerful paradigm for tackling many real life challenges in the space of NP-hard combinatorial problems. While many consumer grade implementations of constraint solvers are available, the processes of correctly modelling a problem, as well as choosing and configuring a suitable solver remain an art usually reserved for experts. In this paper we outline a PhD research project aimed to reduce expert knowledge in and improve performance by Constraint Satisfaction/Optimization Problem Transformation and Constraint Solver Configuration. The paper describes the problems, poses research questions, proposes experiments, summarizes the related work and presents the current experimental progress.

Keywords: Constraint programming · MiniZinc · Machine learning

1 Problem Statement

Constraint Programming (CP) [3] is a powerful paradigm for tackling many real life challenges. While many consumer grade implementations of constraint solving libraries are available, the process of modelling remains an art. Throughout the process of solving a problem with the means of CP, expert knowledge comes into play at two points: Modelling and transforming a model, and configuring the constraint solver (search strategy & parallelization strategy). Both, the way the problem is modelled, as well as the solver configuration, can have significant influence on how quickly a (good) solution can be found. It is desirable to automate and optimize the process for both steps, so that neither expert knowledge is necessary, nor may human error lead to suboptimal configuration decisions.

2 Research Questions

From the stated problem a selection of research questions arise. The first set of questions stems from the modelling aspect of CP. We assume that an initial

J. M. Machado et al. (Eds.): DCAI 2022, LNNS 585, pp. 177–183, 2023.
https://doi.org/10.1007/978-3-031-23210-7_19

model is given by the user who wishes to apply CP to solve a problem. From here a model can potentially improved with regards to the time it takes to solve it or to find an optimal solution by transforming the complete or parts of the problem.

So the first set of questions looks at whether the optimization of constraint models by transformation can be guided based on machine learning. For many constraint model transformations it can be observed that in some cases the transformation leads to an improvement in resolution speed (on the given hardware) and in some cases it does not. Now this must be due to either a change in the resulting search tree (given a search configuration), a change in effort needed for constraint propagation or a difference in performance on the given hardware.

The arising questions here are:

1. Which transformations are available for constraint models and qualify as potentially performance increasing in a subset of cases?
2. Which attributes of constraint models are significant indicators whether to apply a transformation or not?
3. Which transformations are to be applied, given a set of possible transformations and how do these influence each other?

Another area of interest, following a readily modelled and possibly transformed CSP is the selection of an optimal search strategy and possibly a parallelization configuration. Here again similar questions arise:

4. Can one learn to choose an optimal constraint solver, search strategy and/or parallelization approach?
5. Which are significant features for learning and how does the given hardware influence these decisions?
6. How do both decisions influence each other, can one combine them and how may they be influenced by the decisions taken earlier in the pipeline?

3 Related Work

Analysing optimisation problems in NP in order to better understand them and to possibly utilise statistical methods or machine learning has received research interest both recently as well as back when machine learning was not a widely available option. Since there is a lot of research available to cover, we focus on a small number of select publications here. A more detailed analysis of the state of the art can be found in [6].

In [2] Cheeseman et al. identify so-called "order parameters" for a number of specific problems, suggesting these exist for every/most problems. These manually identified order parameters correlate with the time needed to solve the problem examined in the paper and suggest that there are quantifiable attributes that are characteristic for hard to solve problems, and that hard instances or problems are not unlucky outliers. Liu also detailed this in [8, Section 5.1] for the Social Golfer Problem where the order parameters are not obvious and arise from details of the Social Golfer Problem's structure.

Gent et al. [4] take a specific look at utilising nogood learning in constraint solving as a technique that does not always pay off. They identify a subset of problem features that they consider to reflect "the structure of a constraint problem", and also prune the whole set of features for ones that are easy to compute or that turn out to be important by parameter optimization. They then use these to build fast and cheap classificators that perform well in deciding whether their lazy nogood learning is a worthy algorithmic investment (on average) for a problem. The same authors utilize the same feature set in [5] to create a classifier which on average successfully decides for one of nine different *alldifferent* propagator implementations.

In [11] Nudelman et al. investigate two different SAT distributions to find metrics that correlate with instance hardness besides the previously popularized clauses-to-variables ratio. They create and analyze variable and clause-based graphs, as well as doing local probes with randomly partially instantiated variables and local search algorithms. While they had great success creating solver portfolios with their "hardness models", their features are near exclusively unique to SAT problems and don't directly translate to more generic constraint problems.

In [6,7] Hutter et al. give a comprehensive summary on the state of the art of algorithm runtime predictions and the used features, as well as proposing some new ones. They focus on SAT, TSP and MIP problems. To this end they provide lists of features previously and newly described specifically for SAT, MIP and TSP problems. The lists include both rather general features, such as features related to graph representations, as well as features that are specific to the examined problems, such as clause learning features (SAT) or attributes based on a minimum spanning tree created for the TSP problem. Further features are derived from the computational cost of extracting features on a graph, as well as statistics from running solvers for a limited amount of time. While they utilize a number of statistical and machine learning tools, they achieve either the best, or close to the best results with their predictions with random forests.

Our work intends to examine these described approaches on a more general body of constraint problems, rather than focussing on instance collections of a select few problems. Furthermore we aim to extend this to model transformation as well as solver configuration.

4 Proposal

We propose a number of experiments aimed at answering the previously stated research questions. First we aim to provide the necessary infrastructure to systematically analyze Constraint Satisfaction and Constraint Optimisation Problems (CSPs and COPs). In order to leverage the research communities efforts in sourcing interesting and benchmarkable problems, we focus on analysing problem models noted in the MiniZinc Modelling Language [10]. Therefore we need a parser that allows us to read MiniZinc files for further analysis. This analysis aims to focus on two aspects: First the representation of a CSP in various graph forms, and secondly the analysis of CSP specific attributes of the models represented by looking at model semantics as well as attributes of the MiniZinc abstract syntax tree (AST).

The results of these analyses, besides providing interesting insights into the researched body of CP models, can then be leveraged as features for statistical methods/machine learning looking to make meaningful predictions with regards to model transformations and solver configuration. They should also allow to reason on the differences between models and why some models profit from certain transformations or solver configurations while others do not.

As should become evident from Sect. 3 most research into identifying key attributes of CSPs or optimizing solver or portfolio configuration focuses on only a few models—but many instances of these—at a time. We hope to gather insights more broadly applicable by focusing on a broader body of semi-realistic problems, rather than only analyzing a few problems or generated models. Therefore we propose to source the available catalogue of CSPs and COPs collected through the many years of staging The MiniZinc Competition [13]. We would furthermore like to include some of the problems presented in the CSPLib [1].

We then aim to apply different model transformation on the gathered CSPs and benchmark these together with the original models. The results from these experiments should allow us to utilize statistical and machine learning methods to provide predictors that reliably tell whether a transformation is worth the conversion effort.

The same applies for solver configuration: Benchmarking various solver configurations on the same model should provide an insight on which model features correlate with certain search techniques being advantageous and might allow to speed up average solving times by automatically choosing a suitable solver configuration.

5 Preliminary Results

In our current experiments we source and analyze models from The MiniZinc Competition, in total 778 problem instances from the years 2013–2022, and correlate these with the raw results achieved within these competitions. This aims to be a preliminary stage, testing our model analysis capabilities on historic data, before implementing and benchmarking model transformations. This also allows to test our analysis on a simplified version of our algorithm selection, choosing the correct solver for a problem.

We parse these problems and generate binary graphs from the models. For the analysis we extract a subset of features proven successful in previous performance prediction [6,7] and algorithm selection experiments [4]. We extract a number of features, either directly from the AST or the generated binary graph. The calculated features are:

- variable, decision variable, parameter, predicate and constraint count
- mean constraint arity, mean constraints per variable, normalized mean constraints per variable

- edge density, clustering coefficient (sampled on a randomly selected 500 vertices subset), mean and median normalized vertex degree, standard deviation of normalized vertex degree (normalized by total number of graph vertices)
- size of and time to compute for the AST and the binary graph and the FlatZinc file line count

We then utilized Machine Learning techniques based on the `scikit-learn` framework [12]. All results discussed here utilize either their RandomForestClassifier or -Regressor. We create two data sets: One is based on grouping results by the problem instance and summarizing result values for each instance. It therefore contains 778 data points. The second one maintains all results and can be used to evaluate per-solver performance predictions. This data set spans 24808 data points. All models were evaluated utilizing 5-fold cross validation. We then attempt to predict per problem instance as well as per-run results, such as:

- mean time across all solvers to solve a problem
- winning solvers
- whether any solver is capable of solving the problem
- a solvers expected run time
- a solver is likely to solve a satisfiability instance

to varying degrees of success. Among the better performing predictors were predicting the mean time that solvers need to solve a problem or weather any solver will solve an instance. Predicting winners on the other hand turned out to be difficult. We also tried predicting run times for the Gecode as well as the Choco solver. We focused on these two, as they provide many result data points across the competition years. It was more difficult to predict the run times for the Gecode solver, as it was for the Choco solver, despite both providing many competition results. This might be the result of hardware and software developments across the years of the collected results. The resulting run times are therefore not really normalized. We therefore aim to recreate benchmark times across the whole instance set, rather than relying on the historic competition results, in the future.

Besides creating a uniform data base by rerunning the current solver versions on uniform hardware, we further hope to achieve better results by extracting more CSP model specific attributes. We hope to find specifically helpful information in attributes related to the use of global constraints and the resulting hypergraph. Of further interest would be features that take into account runtime statistics, such as (solver dependent) search metrics with regards to resulting search node counts, backtrack counts or learned implicit constraints (nogoods). The search tree based metrics are also relevant when judging a models difficulty more independently from the used hardware. With this infrastructure in place we hope to create good predictors for model transformation payoff.

Acknowledgements. We would like to thank the organizers of the MiniZinc Challenge both for organizing it, as well as providing all submitted problems and results open access. We would also like to thank Hans Würfel, Michael Marte as well as the scikit-learn contributors for the tools they provide that made this research so far possible [9, 13, 14].

References

1. CSPLib: A Problem Library for Constraints. http://www.csplib.org (1999)
2. Cheeseman, P.C., Kanefsky, B., Taylor, W.M.: Where the really hard problems are. In: Mylopoulos, J., Reiter, R. (eds.) Proceedings of the 12th International Joint Conference on Artificial Intelligence, Sydney, Australia, 24–30 Aug 1991, pp. 331–340. Morgan Kaufmann (1991)
3. Dechter, R.: Constraint Processing. Elsevier Morgan Kaufmann (2003)
4. Gent, I.P., Jefferson, C., Kotthoff, L., Miguel, I., Moore, N.C.A., Nightingale, P., Petrie, K.E.: Learning when to use lazy learning in constraint solving. In: Coelho, H., Studer, R., Wooldridge, M.J. (eds.) ECAI 2010–19th European Conference on Artificial Intelligence, Lisbon, Portugal, 16–20 Aug 2010, Proceedings. Frontiers in Artificial Intelligence and Applications, vol. 215, pp. 873–878. IOS Press (2010)
5. Gent, I.P., Kotthoff, L., Miguel, I., Nightingale, P.: Machine Learning for Constraint Solver Design—A Case Study for the All Different Constraint. CoRR abs/1008.4326 (2010). http://arxiv.org/abs/1008.4326
6. Hutter, F., Xu, L., Hoos, H.H., Leyton-Brown, K.: Algorithm runtime prediction: methods & evaluation. Artif. Intell. **206**, 79–111 (2014). https://doi.org/10.1016/j.artint.2013.10.003
7. Hutter, F., Xu, L., Hoos, H.H., Leyton-Brown, K.: Algorithm runtime prediction: methods and evaluation (extended abstract). In: Yang, Q., Wooldridge, M.J. (eds.) Proceedings of the Twenty-Fourth International Joint Conference on Artificial Intelligence, IJCAI 2015, Buenos Aires, Argentina, 25–31 July 2015, pp. 4197–4201. AAAI Press (2015). http://ijcai.org/Abstract/15/595
8. Liu, K.: Parallel Constraint Solving for Combinatorial Problems. Ph.D. thesis, Brandenburg University of Technology, Cottbus, Germany (2021). https://opus4.kobv.de/opus4-btu/frontdoor/index/index/docId/5437
9. Marte, M.: Minizinc-Challenge-Results. https://github.com/informarte/minizinc-challenge-results (2020)
10. Nethercote, N., Stuckey, P.J., Becket, R., Brand, S., Duck, G.J., Tack, G.: Minizinc: towards a standard CP modelling language. In: Bessiere, C. (ed.) Principles and Practice of Constraint Programming - CP 2007, 13th International Conference, CP 2007, Providence, RI, USA, 23–27 Sept 2007, Proceedings. Lecture Notes in Computer Science, vol. 4741, pp. 529–543. Springer, Heidelberg (2007)
11. Nudelman, E., Leyton-Brown, K., Hoos, H.H., Devkar, A., Shoham, Y.: Understanding random SAT: beyond the clauses-to-variables ratio. In: Wallace, M. (ed.) Principles and Practice of Constraint Programming - CP 2004, 10th International Conference, CP 2004, Toronto, Canada, September 27–October 1, 2004, Proceedings. Lecture Notes in Computer Science, vol. 3258, pp. 438–452. Springer, Heidelberg (2004)
12. Pedregosa, F., Varoquaux, G., Gramfort, A., Michel, V., Thirion, B., Grisel, O., Blondel, M., Prettenhofer, P., Weiss, R., Dubourg, V., Vanderplas, J., Passos, A., Cournapeau, D., Brucher, M., Perrot, M., Duchesnay, E.: Scikit-learn: machine learning in python. J. Mach. Learn. Res. **12**, 2825–2830 (2011)

13. Stuckey, P.J., Becket, R., Fischer, J.: Philosophy of the MiniZinc challenge. Constraints Int. J. **15**(3), 307–316 (2010)
14. Würfel, H.: FlatZincParser.jl. https://github.com/hexaeder/FlatZincParser.jl (2021)

Check for updates

The Impact of Covid-19 on Student Mental Health and Online Learning Experience

Faiz Hayat[1](✉), Ella Haig[1], and Safwan Shatnawi[2]

[1] School of Computing, University of Portsmouth, Portsmouth PO1 3HE, Hampshire, UK
{faiz.hayat,ella.haig}@port.ac.uk

[2] College of Applied Science, University of Bahrain, Shakeer, Bahrain

Abstract. The ongoing coronavirus pandemic has affected every facet of human life in the contemporary world. Consequently, university students have to adjust to radically change learning environments. Moreover, the movement restrictions from the government-imposed lockdowns negatively affected students' mental health due to mental issues such as stress, frustration, and depression. The pandemic has caused considerable changes in our daily lives. These reasons are why the virus has hurt individuals' mental health, especially students who had to cope with changes in the education system and even the loss of loved ones. The ambiguity resulting from the pandemic has yet to be fully covered, particularly the students' well-being and the new learning landscape that they are anticipated to navigate seamlessly without their usual support systems. Covid-19 did disrupt the normal and put us all in numerous stressful circumstances' and forced us to have to face overwhelming difficulties at a time. Covid-19 lockdown and pandemic did bring about a sense of anxiety and fear around the world. The spectacle has led to students' long-term and short-term mental health and psychological implications. The paper presents research showing that most students were not prepared for this change, and that indeed they were affected mentally by remote learning. Additionally, the effect of prolonged pandemic fatigue and lockdown on university scholars and academic experiences is unclear. This paper reviews articles about mental health aspects of students and online learning experiences impacted by Covid-19 and provides a roadmap for an ongoing research.

Keywords: COVID-19 · Mental health · Remote learning · Pandemic · Lockdown

1 Introduction

The coronavirus pandemic forced universities to postpone or cancel physical learning on campus instead of utilizing remote learning to teach [1]. Moreover, shifting from previous support systems that consisted of family and friends can be particularly overwhelming [2]. The pandemic changed the way society functions daily, from face-to-face learning format to online learning or remote learning. Besides, developing new group relations can be very intricate for potential learners, resulting in phases of solitude in students, and has been significantly associated with enhanced stress, depression, and anxiety [3].

J. M. Machado et al. (Eds.): DCAI 2022, LNNS 585, pp. 185–190, 2023.
https://doi.org/10.1007/978-3-031-23210-7_20

Therefore, there is a pressing need to analyze the coronavirus pandemic's outcomes on undergraduates' well-being by assessing the findings from peer-reviewed literature and studies [4].

With the pandemic, different schools had to close immediately. Findings from the research show that the general public did experience psychological and mental health impacts of the Covid-19 outbreak, like worrying about others, anxiety, and fear. Longitudinal studies did show that compared to previous academic terms, people in the winter 2020 term seemed to be more depressed, anxious, and passive. Additionally, different behaviors, like increased use of phones, fewer location visits, and lower physical activities, were associated with fluxes in Covid-19 news reportage. Some findings linking students' fears regarding Covid-19 were associated with the place of residence, sources of parent's revenue, whether living with maternities, and whether an acquaintance or relative was impacted by the pandemic [5]. Some scholars are at higher risk of the dangers of societal separation and growth of psychological issues in the pandemic.

The rest of the paper is organized as follows. Section 2 outlines the related work by other researchers. Section 3 presents the primary research questions, research methodology and analysis is mentioned in Sect. 2. Finally, Sect. 5 concludes the paper and outlines directions for future work.

2 Literature Review

The paper will get into an in-depth understanding of the impact of Covid-19 on students from the existing literature.

The research revealed that a considerable section of the student population is at increased risk of psychological outcomes during the coronavirus outbreak [16]. However, the study indicated that gender was not liked with Covid-19 associated stressors. Thus, the research is incoherent with a Chinese study implying that males indicate considerable depression, stress, and anxiety levels. Additionally, it is difficult to assert that house quarantine results in depression since DASS 21 scale is independent of occurrences. However, the researchers assert that assistance from mental health practitioners is imperative. However, the study did not consider students' perception of remote learning and its effect on learners' mental welfare.

The study offers a preliminary understanding of mental well-being and connected behaviors during the pandemic's initial stage. The research revealed increased depression and anxiety during the 2020 winter semester. The trend seemed to heighten as students approached the examination period [6]. Moreover, the study indicated high depression and stress rates during the onset of the pandemic resulting increased phone use, limited movement [7]. However, research gaps exist regarding screen types used by students during the pandemic period and quantify the amount of information gathered from social media and digital news outlets.

Remote learning presents stumbling blocks such as low internet speeds, which affect the students' learning experience [8]. Besides, the ambiguity regarding university assessment enhances anxiety levels within learners. Moreover, the doubt regarding finishing assessments by utilizing new techniques enhances students' anxiety levels, particularly when they feel that the new methods may fail to capture their real abilities [10]. The research reveals disruption of physical learning in higher learning institutions affected counseling services [15]. Offering well-being and counseling sessions are significant elements of students' support [13]. They can access different extents of support for severe psychological health problems and more persistent, long-term conditions [12]. The study outlined various coping mechanisms diverse between adaptive and maladaptive activities [11]. Dysfunctional coping mechanisms like denial and detachment have significantly predicted depression among university and college students [14]. However, the results from these researches are inadequate, diverse, and carried out within a surfeit of various circumstances.

The findings outline the extent of difficulties in remote learning, increased psychological problems, and stumbling blocks to dealing with mental health issues during the pandemic. Thus, the findings provide a rationale for discussions to delve deeply into the issues and offer feasible solutions.

3 The Primary Research Questions

The aim of the research is to investigate the impact of the pandemic (Covid-19) on university students' mental health and education.

The research questions are formulated as follows:

1. What is the current students' mental health status and their experience with education delivered under the constraints of Covid-19?
2. What are the stressors, which lead to stress, anxiety and depression among the university students?
3. Is there any relation between university students' mental health and social media usage during the pandemic?
4. Is there any relation between university students' remote learning experience and social media usage during the pandemic?
5. What are the experiences, issues and challenges faced by university students in the context of mental health during the pandemic?
6. What are the experiences, issues and challenges faced by university students in the context of remote learning during the pandemic?

4 Methodology

4.1 Research Methods and Data Analysis

The study will use qualitative and quantitative methods to gather information. The method used for the research is Questionnaire. The method is used by other researcher in their studies to understand the student's current situation such as [5, 6, 9]. Questionnaire is an integral part of the research which will gather information on the university students' current well-being by asking questions related to their stress level, analysing stressors, and adapting to the remote learning setting. The university students selected for the study include all the possible type i.e. undergraduate, postgraduate, full-time and part-time, international university students currently in the country of study or abroad.

We are capturing all the sub categories mentioned above so we can compare between different student groups and capture any variation between them.

Procedure for Data Collection

An online survey will be used to collect responses using Microsoft Forms. The survey will be distributed among various university students from several countries. Different University administrators will be contacted to distribute the survey. An email will be sent to everyone on the mailing lists by way of a distributor inviting them to participant in the survey.

Data Analysis

The next phase of the research includes application of various tools on the collected data for Data Analysis. Depending on the nature of the data various tools and methods (such as the SPSS software and the ANOVA and t-test methods) will be considered.

5 Conclusions and Future Work

Many individuals' lives have been impacted by the Covid-19 epidemic worldwide. The fast rise of disease cases worldwide has produced fear and uncertainty about what will occur next. It has also resulted in a great deal of anxiety among children. Previous research has demonstrated that public health catastrophes like the Covid-19 outbreak can have many psychological impacts on university students, including worry, fear, and anxiety. The Covid-19 epidemic has impacted and will remain to have an impact on how skills and knowledge are delivered at all stages of education. Even though many adults and children will adjust to new compensation and modalities to disrupt conventional teaching offerings, some may struggle. The gap between individuals whose households cannot assimilate the teaching and monitoring of qualifications needed for in-home schooling due to a lack of time and abilities is now unaddressed. This frustration may have negative consequences for pupils' academic performance and mental health.

While not intended to be comprehensive, the poll tried to learn more about the effect of the Covid-19 epidemic on university students' mental health and education [5]. The poll received $n = 557$ answers from university students at the Lucerne University of Applied Sciences and Arts who are enrolled in the institution's from all the six departments. In terms of stress and worry, the study found that 85.8% of the undergraduates had feelings of stress; however, most of these signs were minor (63.3%). The research did not back up earlier results that scholars who live solitary are more likely to suffer mental health issues. The research did not corroborate prior results indicating scholars who live solo are more likely to suffer mental health issues. On the other hand, female scholars seemed to be at a greater risk of unfavorable mental health outcomes, although the impact was tiny. We cannot tell if these signs occurred even before the epidemic because this research is cross-sectional. However, many probable reasons students would be anxious or nervous during the Covid-19 epidemic notably increased study-related issues or concerns about employment prospects.

Nonetheless, the study's findings imply that the students dealt very well with the anxiety during the shutdown. Furthermore, most of the scholars felt properly supported and expressed gratitude to the professors. However, neither the university administration nor the teachers should rest on their merits. The answers to the open-ended quiz show that instructor found online teaching difficult, which caused scholars to get stressed. As a result, maybe more than stress and anxiety, the experiences of a swift digital transition to distant instruction have exposed much about the university education sector's weaknesses and, therefore, much of what needs fixing in institutions. Students and lecturers alike must be ready for future periods that will necessitate adaptability, a considerable workload, and more learning activity. Online learning is no longer a "nice to have" for both students and instructors" but a necessary skill. There are numerous grounds to suspect that Covid-19 has established a "new normal" for colleges, one which will persist once the shutdown is lifted. The fast development of ICT and the increased complexity accompany its immense potential reasons why digitalization in education remains to attract special attention, especially in the aftermath of the Covid-19 outbreak. It is the responsibility of the university administration to provide the essential tools for both students and instructors to gain these skills.

References

1. Pragholapati, A.: COVID-19 impact on students (2020). https://doi.org/10.17605/OSF.IO/NUYJ9
2. Mseleku, Z.: A literature review of e-learning and e-teaching in the era of Covid-19 pandemic. SAGE **57**(52), 588–597 (2020)
3. Chirikov, I., Soria, K.M., Horgos, B., Jones-White, D.: Undergraduate and graduate students' mental health during the COVID-19 pandemic. Center for Studies in Higher Education, UC Berkeley (2020). Retrieved from https://escholarship.org/uc/item/80k5d5hw
4. Savitsky, B., Findling, Y., Ereli, A., Hendel, T.: Anxiety and coping strategies among nursing students during the covid-19 pandemic. Nurse Educ. Pract. **46**, 102809 (2020)
5. Lischer, S., Safi, N., Dickson, C.: Remote learning and students' mental health during the Covid-19 pandemic: a mixed-method inquiry. Prospects 1–11 (2021, online first)
6. Huckins, J.F., DaSilva, A.W., Wang, W., Hedlund, E., Rogers, C., Nepal, S.K., Wu, J., Obuchi, M., Murphy, E.I., Meyer, M.L., Wagner, D.D., Holtzheimer, P.E., Campbell, A.T.: Mental health and behavior of college students during the early phases of the COVID-19 pandemic: longitudinal smartphone and ecological momentary assessment study. J. Med. Internet Res. **22**(6), e20185 (2020)
7. Chaturvedi, K., Vishwakarma, D.K., Singh, N.: COVID-19 and its impact on education, social life and mental health of students: a survey. Child Youth Serv. Rev. **121**, 105866 (2021)
8. Al-Taweel, D., et al.: Multidisciplinary academic perspectives during the COVID-19 pandemic. Int. J. Health Plann. Manage. **35**(6), 1295–1301 (2020)
9. Zhai, Y., Du, X.: Addressing collegiate mental health amid COVID-19 pandemic. Psychiatry Res. **288**, 113003 (2020)
10. Mishra, L., Gupta, T., Shree, A.: Online teaching-learning in higher education during lockdown period of COVID-19 pandemic. Int. J. Educ. Res. Open **1**, 100012 (2020)
11. Son, C., Hegde, S., Smith, A., Wang, X., Sasangohar, F.: Effects of COVID-19 on college students' mental health in the United States: an interview survey study. J. Med. Internet Res. **22**(9), e21279 (2020)
12. Tasso, A.F., Hisli Sahin, N., San Roman, G.J.: COVID-19 disruption on college students: academic and socioemotional implications. Psychol. Trauma Theory Res. Pract. Policy **13**(1), 9 (2021)

13. Vijayan, R.: Teaching and learning during the COVID-19 pandemic: a topic modeling study. Educ. Sci. **11**(7), 347 (2021)
14. Rogowska, A.M., Pavlova, I., Kuśnierz, C., Ochnik, D., Bodnar, I., Petrytsa, P.: Does physical activity matter for the mental health of university students during the COVID-19 pandemic? J. Clin. Med. **9**(11), 3494 (2020)
15. Burns, D., Dagnall, N., Holt, M.: Assessing the impact of the COVID-19 pandemic on student well-being at universities in the United Kingdom: a conceptual analysis. In: Frontiers in Education, vol. 5, p. 204. Frontiers (2020, Oct)
16. Khan, K.S., Mamun, M.A., Griffiths, M.D., et al.: The mental health impact of the COVID-19 pandemic across different cohorts. Int. J. Ment. Health Addict. **20,** 380–386 (2022). https://doi.org/10.1007/s11469-020-00367-0

Threat Detection in URLs by Applying Machine Learning Algorithms*

Álvaro Bustos-Tabernero, Daniel López-Sánchez, and Angélica González Arrieta(✉)

University of Salamanca, Plaza de los Caídos, 37008 Salamanca, Spain
{alvarob97,lope,angelica}@usal.es

Abstract. Different cyber threat groups develop an infrastructure to be able to distribute malware to victims. The entry vector of these threats is usually the download of malicious files via a web link that initiates the system infection. It is possible to detect these threats at an early stage to anticipate a possible compromise by applying a malicious URL detector. This work contributes to detect cyber threats (Emotet and Qakbot, mainly) in advance given an input URL.

Keywords: Malware detection · Cyber intelligence · Threat hunting · Machine learning · Security

1 Problem Statement

Security can be defined as a constantly shifting search for a balance between attackers and defenses. With the creation of new technologies, attackers tend to exploit new avenues and explore new tactics. On the contrary, defenders adapt reactively to these new scenarios [18]. If a study of the different Techniques, Tactics and Procedures (TTPs) were to be carried out, we would be able to abstract the possible movements of cybergroups. These TTPs are descriptions of the behavior of an actor, a group, or a threat. Tactics describe such behavior at a much higher level, while techniques provide more description in the context of a tactic. In contrast, procedures are at a lower level of detail and more detailed description under the context of a technique [6].

This knowledge about the threat can be distributed to all layers of defenses, reducing the gap between advanced attacks and organizational defenses. This discipline is commonly referred to as *Threat Intelligence* (TI) [3]. *Threat intelligence* or *cyber threat intelligence* is the "evidence-based knowledge, including context, mechanisms, indicators, implications and practical advice about a threat, existing or emerging threat to assets that can be used to inform decisions about the subject's response to that threat or threat" [11]. The information gathered enables risk assessment of the threat and decision making accordingly.

* University of Salamanca

J. M. Machado et al. (Eds.): DCAI 2022, LNNS 585, pp. 191–196, 2023.
https://doi.org/10.1007/978-3-031-23210-7_21

1.1 Hypothesis

Certain security measures are taken reactively in the face of a known threat. That is, intelligence services look to public sources for Indicators of Compromise (IoCs) that infect the organization's systems. The process of collecting all the IoCs is quite costly. The whole list of URLs ends up being blacklisted in some web browsing security devices, such as proxy servers. To address this situation, we propose the creation of a machine learning model that detects malware download URLs in real-time.

2 Related Work

From the early stages of *cyber kill chain* [15] (reconnaissance, weaponization, and delivery), it is possible to recognize the patterns of the malicious infrastructure to avoid the first infection phase.

There are several models that are used for phishing detection applying different algorithms [8]. Based on what has been studied for these types of social engineering attacks, these models can be used to detect new, more sophisticated threats.

In the survey [8], the authors summarize several algorithms (support vector machines, bayesian network, decision trees, etc.) in combination with feature extraction from URL or websites. Most of them, they take into consideration words or HTML features [19,20]. On the contrary, the paper [16] applies different natural language processing techniques to analyze the semantics of the website. It is even possible to extract the semantic components of the corpus by applying basic transformers such as TF-IDF [17] or it is even possible to extract sentimental characteristics from web content [2]. Nevertheless, it is too complicated to analyze the payload of a malicious website or URL, due to the fact that it can be downloaded from any kind of file. Hence, the only study object should be the web link.

2.1 Word Embedding

In search of a more sophisticated model, it is possible to apply word embedding techniques such as *Word2Vec* [12], where each of the elements of the URLs is transformed into numeric vectors respecting the similarity between the different tokens and their context. This proposal is based on words and a vocabulary, but it is not prepared for links. Therefore, in the article [22], an adaptation of the previous model is shown.

Given a URL u, it can be separated into several segments in which we will denote $u = (S_1 S_2 S_3 S_4 S_5)$; where S_1 is the protocol or the schema of the link; the subdomain is S_2; S_3 is the domain, and S_4 the domain suffix. Finally, S_5 is the URL path.

Each element S_i is composed by $S_i = (c_1, c_2, \ldots, c_n)$, where c_i is a character and n is an unfixed length of the part [22]. While in previous models of natural language processing, the vocabulary is the set of English words, in this case, the vocabulary is the set of characters used in a URL. Then, to obtain the embedding for the characters, SkipGram language model [10] is used. The SkipGram tries to maximize the co-occurrence probability of words that appear in a window w in a sentence [12]. For instance, an URL u, and a section S_i, for each $c_i \in S_i$ and $c_k \in u[i-w : i+w]$, we maximize the posterior probability of its neighbors [22].

Given the embedded vector of the Skipgrams result of each S_i, we perform an average operation of each of the feature vectors obtained. Each embedding vector of length N. Finally, each vector of S_i is concatenated and the final array has $5N$ length. After conversion, different classification algorithms are applied to determine the nature of the URL [22].

In case [9], a more rigorous classification than the previous cases is elaborated, where Convolutional Neural Networks (CNN) are used. This new proposal addresses the limitations of some previous models such as the inability to obtain information from previously unseen words. CNNs have achieved good results in property extractions in images. In this case, by applying one-dimensional CNNs, information is obtained from two types of features: words and characters. In a URL, words are divided by special characters and punctuation marks.

This neural architecture is mainly composed of two branches: the one that extracts the features of the characters in a the URL corpus, and the branch that analyzes the words. The advances presented in this model allow a detailed analysis of the properties of a URL, beyond the words and their position. In addition, by adding the extraction of character features, it is possible to better management decisions in the face of rare or pseudo-random strings, often used by malicious URLs [9].

3 Proposal

URLNet [9] is capable of managing previously unseen tokens and contextualizing them. This allows us to detect sophisticated patterns that are used by different cyber groups that distribute malware.

On the one hand, we are going to focus on the detection of two threats that are being distributed on a large scale: qakbot and emotet. Qakbot is a trojan virus that it has been used by financially-motivated actors, but it has been evolved into a delivery agent for ransomware [14]. On the other hand, Emotet malware is primarily used as a downloader for other malware variants [13]. Typically, the different actors usually deliver this malware to victims via email. There are some variants where the malicious email has an attached office document that automatically accesses a website to download the malware and trigger the infection of the system or company [1].

For the organization's cyber intelligence services, it is possible to collect the URLs from public sources that are associated with these threats and use them as a training dataset for our network. We have modified some layers of the neural network to be able to classify the different threat types. The final architecture is shown in Fig. 1.

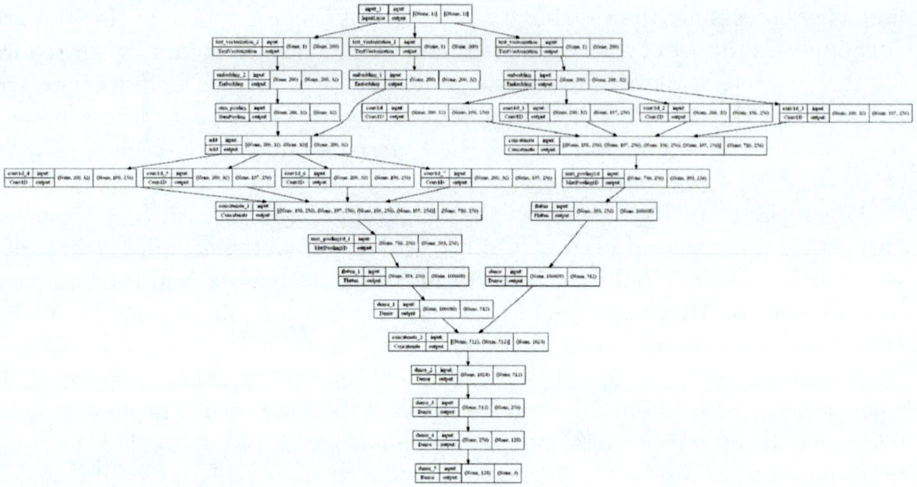

Fig. 1. URLNet adaptation.

With this new model added to some element of web protection, it is possible to detect these threats and block their download, thus preventing infection of the system or organization. For instance, a Web Application Firewall (WAF) is a specific firewall that monitors HTTP traffic by detecting certain web attack patterns and blocking them if necessary. In our case, it is possible to add a new module with the presented model and block those attack patterns that it detects as a threat.

3.1 Model Configuration

This model is trained using the ADAM optimizer [7] with a learning rate value $\alpha = 0.001$, an exponential decay rate $\beta_1 = 0.9$ for the first moment estimates, and $\beta_2 = 0.999$ for the second moment. We have used cross-entropy loss function for the network training.

4 Preliminary Results

For the first results and analysis of the model, we proceeded to create a dataset with benign URLs from the source [4] and the malicious URLs of emotet and qakbot from the URLhaus website [21].

The dataset is made up of $N = 18896$ elements $\{(u_1, y_1), \ldots, (u_N, y_N)\}$, where u_i is a web link and y_i is the label of that URL. In this case, $y_i \in \{0, 1, 2\}$, in which 0 means an emotet URL, 1 a benign or neutral link, and 2 is a qakbot site. For the training phase, 70% of the complete dataset will be used, where 20% of it will be the validation dataset. The remaining 30% will be used for the testing phase.

Once the training has been performed, it has been observed that the accuracy obtained is 98.01%. Although the errors in the model evaluation that we can see in Fig. 2 are small, it is important to realize that the bad categorizations are around label 2 (neutral/benign). It is true that there are false positives and false negatives around this label. The reason for this misclassification will be studied in the future. Conversely, note that it is possible that several malicious files could be distributed by the same actors, as discovered in early 2022 by this security researcher [5].

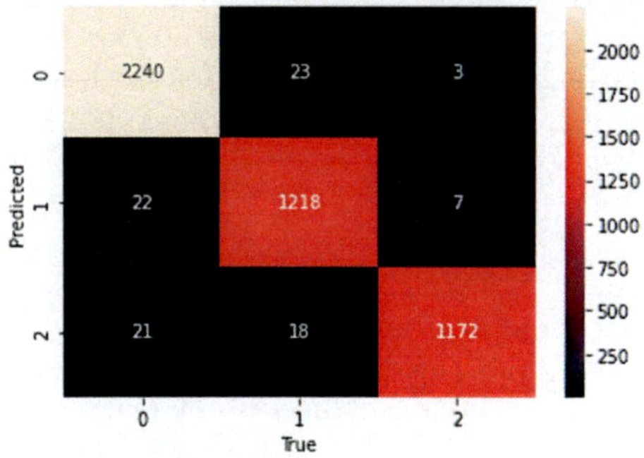

Fig. 2. Confusion matrix.

5 Reflections

With the proposed architecture, we have taken a first step towards detecting threats that spread through malware downloads given a URL. This model can perform an analysis of the patterns in a web link, based on the words, their characters, and the relationship that exist between each of these tokens. This is a great improvement over URLs with pseudo-random strings. Although progress has been made in real-time detection of threats, it is still possible to stay ahead of them. For future research, we intend to develop a threat URL generator based on an architecture similar to the one proposed in order to be able to anticipate the patterns followed by the different actors.

Acknowledgements. This work has been supported by the project "XAI - XAI - Sistemas Inteligentes Auto Explicativos creados con Módulos de Mezcla de Expertos", ID SA082P20, financed by Junta Castilla y León, Consejería de Educación, and FEDER funds.

References

1. Certified Information Systems Auditor (CISA): Qbot/Qakbot Malware Report. Oct. 29, 2020. https://www.cisa.gov/stopransomware/qbotqakbot-malware-report (visited on 08 May 2022)

2. Basarslan, M.S., Kayaalp, F.: Sentiment analysis with machine learning methods on social media **9**, 5–15 (2020). https://doi.org/10.14201/ADCAIJ202093515
3. UK-CERT: An introduction to threat intelligence (2014). https://www.ncsc.gov.uk/files/An-introduction-to-threat-intelligence.pdf (visited on 30 Dec 2021)
4. Canadian Institute for Cybersecurity. URL dataset (ISCX-URL2016). http://205.174.165.80/CICDataset/ISCX- URL- 2016/ (visited on 28 Jan 2022)
5. Fernández, G.: Confirmo, los operadores de #Emotet ocupan el mismo proveedor de Webshells que ocupa TR distribution con #Qakbot o #Squirrelwaffle. Jan. 24, 2022. https://twitter.com/1ZRR4H/status/1485413045975330822?s=20&t=cuUfe7pnr7sZ1YEYcqm7uQ (visited on 27 Jan 2022)
6. Johnson, C., et al.: Guide to cyber threat information sharing **150** (2016). https://doi.org/10.6028/NIST.SP.800-150 (visited on 30 Dec 2021)
7. Kingma, D.P., Lei Ba, J.: Adam: a method for stochastic optimization. In: International Conference on Learning Representations, ICLR (2015)
8. Kiruthiga, R., Akila, D.: Phishing websites detection using machine learning. Int. J. Recent Technol. Eng. **8**(2 Special Issue 11 Sept. 2019), 111–114. ISSN: 2277-3878 https://doi.org/10.35940/ijrte.B1018.0982S1119
9. Le, H., et al.: URLNet: learning a URL representation with deep learning for malicious URL detection. ArXiv (May 2018). arxiv.org/abs/1802.03162
10. McCormick, C.: Word2Vec Tutorial - The Skip-Gram Model. In: 2016
11. McMillan, R.: Definition: threat intelligence. In: Gartner Research (2013). https://www.gartner.com/en/documents/2487216 (visited on 30 Dec 2021)
12. Mikolov, T., et al.: Efficient estimation of word representations in vector space (2013). arxiv.org/abs/1301.3781. https://doi.org/10.48550/ARXIV.1301.3781
13. MITREATT& CK. Emotet. https://attack.mitre.org/software/S0367/ (visited on 08 May 2022)
14. MITREATT&CK. QakBot. https://attack.mitre.org/software/S0650/ (visited on 08 May 2022)
15. Nikkhah, P., et al.: Cyber kill chain-based taxonomy of advanced persistent threat actors: analogy of tactics, techniques, and procedures. Threat detection in URLs by applying machine learning algorithms 7. J. Inf. Proc. Syst. **15** (2021). https://doi.org/10.3745/JIPS.03.0126
16. Peng, T., Harris, I., Sawa, Y.: Detecting phishing attacks using natural language processing and machine learning. In: 2018 IEEE 12th International Conference on Semantic Computing (ICSC), pp. 300–301 (2018). https://doi.org/10.1109/ICSC.2018.00056
17. Pimpalkar, A.P., Retna Raj, R.J.: Influence of pre-processing strategies on the performance of ML classifiers exploiting TF-IDF and BOW features **9**, 49–68. https://doi.org/10.14201/ADCAIJ2020924968
18. Schneier, B.: How changing technology affects security. IEEE Secur. Priv. **10**(2), 104–104 (2012). https://doi.org/10.1109/MSP.2012.39
19. J. Shad and S. Sharma. A Novel Machine Learning Approach to Detect Phishing Websites Jaypee Institute of Information Technology. 2018
20. Sönmez, Y., et al.: Phishing web sites features classification based on extreme learning machine. In: 2018 6th International Symposium on Digital Forensic and Security (ISDFS), pp. 1–5 (2018). https://doi.org/10.1109/ISDFS2018.8355342
21. URLhaus. URLhaus - Malware URL exchange. https://urlhaus.abuse.ch/ (visited on 28 Jan 2022)
22. Yuan, H., et al.: URL2Vec: URL modeling with character embeddings for fast and accurate phishing website detection, pp. 265–272 (2018). https://doi.org/10.1109/BDCloud.2018.00050

An Approach to Simulate Malware Propagation in the Internet of Drones

E. E. Maurin Saldaña[1], A. Martín del Rey[2], and A. B. Gil González[3(✉)]

[1] University of Salamanca, 37008 Salamanca, Spain
emaurin@usal.es
[2] Department of Applied Mathematics, IUFFyM, University of Salamanca, 37008 Salamanca, Spain
delrey@usal.es
[3] BISITE Research Group, University of Salamanca, Edificio Multiusos I+D+I, 37007 Salamanca, Spain
abg@usal.es

Abstract. This research addresses the problem of malicious code propagation in a swarm of drones. Its main objective is to establish the conceptual basis and its setting for the modelling framework of malware propagation using mathematical epidemiology as a core. Specifically, this work identifies the disadvantages associated with the traditional mathematical models employed to simulate malware spreading, revealing the most relevant factors to be considered in order to achieve an ad-hoc modeling that better represents the behavior of the propagation of such malicious code in the drone configuration under study. The relevance of this work is both to contribute to a better understanding of how vulnerabilities in these devices could be exploited and to guide the formulation of cybersecurity measures in line with these problems.

Keywords: Malware propagation · Drone swarm · Mathematical epidemiology · Complex networks · Individual-based models

1 Introduction

Currently, the use of Unmanned Aerial Vehicles (UAV) or Remotely Piloted Aircraft (RPA), both for civil and military uses is expanding, and along with it the diversification in their areas of application, such as agriculture, mining, environmental, civil works, aerial surveillance, last mile delivery of retail products, fire control, disaster support, recreational uses, etc. In turn, in [1], the authors have described the role of these devices in the control of the COVID-19 pandemic for surveillance actions, announcements to the population, sanitizing of public places, among others. On the other hand, they have also been used for illicit activities such as entry into prisons, smuggling and entry of prohibited items [2]. In [3], the authors presented several case studies related to the transport of drugs and prohibited substances. In turn, in [4] the author referred to the so-called narco-drones and their crossing between the U.S.-Mexico border. In [5], the authors conducted a study analyzing 19 types of unmanned aerial vehicles and their

J. M. Machado et al. (Eds.): DCAI 2022, LNNS 585, pp. 197–203, 2023.
https://doi.org/10.1007/978-3-031-23210-7_22

various applications, concluding the fact that "drone's technology will be advancing with the current rate of development of technology and it will be more applicable to our daily life in the future".

The use of this type of swarm configuration and its various applications has aroused the interest of several researchers in the area, even giving rise to the term "Internet of Drones (IoD)" [6] in an effort to standardize vocabulary and propose a universal architecture for this type of technology.

This paper focuses on the design of mathematical models to study and analysis of the propagation of malicious code in a swarm of drones. Specifically, the purpose of this work is establishing the conceptual basis on which a framework for modeling this phenomenon could be based, taking mathematical epidemiology as a basis, and thus improve the understanding of the problem and serve as a guide for efficient simulations.

The approach to be followed by this research is focus on different studies cases about the malware evolution that affected several drones' platforms isolated, and then explain the communication protocols and the behavior of malware at swarm drones configuration, explained by using mathematical epidemiology concepts as just mentioned at above paragraph.

The rest of the paper is organized as follows: Sect. 2 is oriented to the basic aspects of cybersecurity, including the most common types of attack in this type of architecture. Section 3 is divided to explain conceptually what is meant by Internet of Drones Networks (IoD), several strategies to deploy swarm drone's configuration, and finally, present some attacks examples scenarios. Section 4 is devoted to introduce the conceptual bases behind the mathematical models that can be associated with the spread of malware in a swarm of drones.

Finally, the conclusions are shown in Sect. 5 as well as the challenges and future work that could complement this topic.

2 Cybersecurity Aspects

The analysis of drone security was reviewed in [7, 8], as well as the cybersecurity methods applicable when assembling AVNs (Aerial Vehicle Networks). In [9], the authors also presented the most common techniques of attacks on this type of technology, with various objectives, such as: degrading, denying, interrupting and/or modifying the information processed through the components of the drone network. This kind of attacks could be done through the following techniques: Denial of Service, Spoofing, Hijacking, Jamming, Others [10, 11].

Among the latter are those of interest for this research, since they deal with attacks by means of Trojans (malware), reverse engineering and deauthentication. Focused now on the state of the art of malware propagation in drone networks, there was some evidence showing successful infection cases, as in 2014, during the cybersecurity conference CODEBLUE, in which the CTO of the company SEWORKS Dongcheol Hong presented the result of the malware infection of an AR.Drone 2.0 device, with the malicious code "HS-DRONE", incorporating in one of its conclusions that such malware could spread from a smart type device or by infecting between drones [12].

On the other hand, there is also evidence of the development of malicious code affecting an UAV, individually and outlining the possibility that such compromise could affect other vehicles within the same network. Specifically, in 2015 was published in the media, the development of the malware called "maldrone" by Rahul Sasi [13], which was of the persistent type (tolerant to equipment reboots) and which took remote control of the victim drone, even blocking the use of the autopilot. In particular, this type of attack exploited a vulnerability in the Parrot AR quadcopter model, allowing the execution of the malicious code as a backdoor, in an ARM Linux operating system, disabling the platform's security and self-control options and allowing a reverse TCP connection to finally take control and allow the attacker to modify the device's actions.

Finally, according to [14], one of the threats described is the possibility of the existence of "malicious drones" or "compromised drones" in an IoD environment, which would affect the integrity of the network or system, giving an example of how a drone can provide false routing information or collude with other drones.

3 Malicious Code Propagation Aspects

3.1 Internet of Drones Networks

According to [15] "An attacker may target the vehicle's on-board systems, the GCS computer, the vehicle's sensors, or the communication between the vehicle and GCS". Indeed in [16] the authors highlight that the Internet of Drones (IoD), is vulnerable to malicious attacks over radio waves frequency space due to the increasing number of attacks and threats to a wide range of security measures for IoD networks. In [17] the author expresses that "UAVs are mainly used on a wireless network, and data traveling over these wireless channels provides an opportunity for cyber attackers to capture sensitive, highly confidential military information. Attackers can plant viruses, Trojans, and other manifestations of malware into the interconnected communication network".

3.2 Attacks Examples Scenarios

A suitable attack example for the mentioned scenario could be feasible by using the Mirai botnet, where the malicious code could be injected into a hijacked UAV. A based case study was shown in [18], where a mobile attacker drone breaks into UAV communication, creating a bot-network, infecting several devices. In that case "the botmaster can issue a command to a peer bot, and the command will propagate through the network".

Another infection vector could be done "infiltrating the design, implementation and production phases to create backdoors in the control software and communications protocols of the drone". In this case, the Mirai source code [19] could be injected to propagate to another neighbor UAVs, then as mentioned in [20], this malware could "exploited default account passwords and hidden backdoors for IoT devices to gain control and carry out a DDoS attack", using device firmware modification as a vector attack, gaining full remote access. Several proposed techniques to detect this kind of attack are discussed in that publication, too. Other techniques, with success in other scenarios, can be applied [21–23].

Focusing now on the Mirai botnet, it has spawned many variants following the same infection strategy that could be adapted to reach the desired goals. For example, in [24] there is a detailed explanation about this botnet structure and propagation through different stages, including a scanning phase, then passthrough brute-force login stage and wait the first successful login to send the victim IP and credentials to a hardcoded report server, while a separate loader program infected these vulnerable devices, downloading and executing specific malware (stage 4). Finally, the attacker could send attack commands from the command and control server, while simultaneously scanning for new victims, according to the publication. Mirai, commonly performs scanning phase looking Telnet service, under ports 23 and 2323, and could be modified to look other ports or services into the victim device, also include HTTPS, FTP, SSH, and others.

4 Conceptual Foundations of the Propagation Model

The proposed framework is based on the possibility of malware infection of drones in swarm configuration, which could affect the aircraft itself (considered as a node), its control station, its communication links, among others, and could spread to other devices in the same network, or other susceptible devices, causing risk to the entire formation.

Our contribution will be oriented to the study of propagation of malicious software (malware) instead of biological agents spreading, and could be based on the work done by Kermack and McKendrick, within the so-called mathematical epidemiology, being its first approach the use of ordinary differential equations to simulate such behaviors, between years 1927 and 1933, and nowadays applicable to various types of simulations and models.

One of these models is the MSEIR type (M: immune class, S: susceptible class, E: exposed class, I: infectious class, R: recovered class) and their derivations according to [25]. These models are mostly global and deterministic type, which do not consider the particular characteristics of each agent, node, or infected equipment (drone, in our case), nor the particular way in which they interact. To solve the disadvantages of such models, there are publications that have referred to the use of cellular automata, agent-based models, being a more realistic approach and covering global, particular issues, in a discrete and stochastic manner, without departing from the advantages of the basic differential equations.

In [26], the authors proposed a mathematical model to simulate the propagation of malware in Wireless Sensor Networks, based on cellular automata and also considered real factors such as the use of intrusion detection devices (IDS), antivirus solutions, autonomy and heterogeneity properties of the agents (nodes), mechanisms used by viruses to propagate, among others, reaching different conclusions for each type of test scenario performed, including its application in complex networks. Based on these preliminary results, it led to the conclusion that it is possible to better simulate the epidemiological state of a certain agent, equipment or device, whether exposed, infected, susceptible, recovered, etc., at a given time, and to predict its behavior both individually and globally, in the case of being isolated or forming part of a network.

In our case, in the malware propagation modeling, the models can be classified according to different characteristics: basically, we have deterministics or stochastics models, continuous or discrete models, and (this is the most important for us) global or individual-based models. Obviously, this classification is not only typical of malware modeling but also of biological disease modeling, and so on.

Finally, our evaluation plan consists in the study of compartmental models where the population of devices can be divided into several classes depending of the role or the state of the device with respect to malware: susceptible, infectious, exposed, recovered, damaged, etc., proposing of the following approaching ways:

- One of them is to reformulate global models dividing the original compartments into subclasses that depend on the number of connections. Thus, continuous models on complex networks are obtained and some characteristics of the network (degree of each node) are considered.
- The other possibility is to consider individual-based models where each agent represents a device and all local interactions are represented by means of a complex network. In this case, we will refer to those models whose dynamics is defined by means of cellular automata.

5 Conclusions and Further Work

Considering the above background and the analysis of the information gathered, it can be concluded that it is convenient, feasible and acceptable to use tools of epidemiological mathematics to model the behavior of malware propagation in UAVs and specifically in this case, for swarm type configurations, which according to the analysis of the cases presented is totally real and feasible. At the same time, the need to adapt the existing classical models (Kermack and McKendrick) for a better approximation to the reality of the behavior of these aerial vehicles is evident. There are aspects that require special treatment for a cybersecurity point of view in a swarm drone, that differs from the classical perspective on land networks, due to the power consumption restraints, the onboard processor capacity, an others factors that must be considering. Also, techniques like agility concepts at ground control station, and based on operational situational awareness panorama could be applied to improve the security levels. Finally, it is necessary for future work to study the applicability of mathematical models based on cellular automata or others, which minimize the disadvantages and consider variables such as the heterogeneity of the agents, their particular characteristics and their network behavior in their various epidemiological states.

Acknowledgments. This research has been supported by the project "Intelligent and sustainable mobility supported by multi-agent systems and edge computing (InEDGEMobility): Towards Sustainable Intelligent Mobility: Blockchain-based framework for IoT Security", Reference: RTI2018-095390-B-C32, financed by the Spanish Ministry of Science, Innovation and Universities (MCIU), the State Research Agency (AEI) and the European Regional Development Fund (FEDER).

References

1. Chamola, V., Hassija, V., Gupta, V., Guizani, M.: A comprehensive review of the covid-19 pandemic and the role of IOT, drones, AI, blockchain, and 5G in managing its impact. IEEE Access **8**, 90225–90265 (2020). https://doi.org/10.1109/ACCESS.2020.2992341

2. Post, T.W.: Prisons Try to Stop Drones from Delivering Drugs, Porn and Cellphones to Inmates, webpage (2016) https://www.washingtonpost.com/local/prisons-try-to-stop-drones-from-delivering-drugs-porn-and-cellphones-to-inmates/2016/10/12/645fb102-800c-11e6-8d0c-fb6c00c90481story.html
3. Turkmen, Z.: A new era for drug trafficking. Drones Forensic Sci. Addict. Res. **2**, 2–3 (2018). https://doi.org/10.31031/fsar.2018.02.000539
4. Schmersahl, A.R.: Fifty Feet Above the Wall: Cartel Drones in the U.S.—Mexico Border Zone Airspace, and What to do About Them. Naval Postgraduate School Thesis (2018)
5. Chan, K.W., Nirmal, U., Cheaw, W.G.: Progress on drone technology and their applications: a comprehensive review. In: Proceedings, vol. 2030, American Institute of Physics Inc., p. 020308 (2018). https://doi.org/10.1063/1.5066949
6. Yahuza, M., et al.: Internet of drones security and privacy issues: taxonomy and open challenges. IEEE Access **9**, 57243–57270 (2021)
7. Lv, Z.: The security of internet of drones. Comput. Commun. **148**, 208–214 (2019). https://doi.org/10.1016/j.comcom.2019.09.018
8. Sedjelmaci, H., Senouci, S.M.: Cyber security methods for aerial vehicle networks: taxonomy, challenges and solution. J. Supercomput. **74**(10), 4928–4944 (2018). https://doi.org/10.1007/s11227-018-2287-8
9. Kotesh, P.: A comprehensive review of unmanned aerial vehicle attacks and neutralization techniques. In: Ad Hoc Networks (2020). https://doi.org/10.1016/j.adhoc.2020.102324
10. Del Rey, M., Ángel, Batista, F.K., Queiruga Dios, A.: Malware propagation in Wireless Sensor Networks: global models vs Individual-based models. ADCAIJ: Adv. Distrib. Comput. Artif. Intell. J. **6**(3), 5–15 (2017). https://doi.org/10.14201/ADCAIJ201763515
11. Sakarkar, G., Kolekar, M.K.H., Paithankar, K., Patil, G., Dutta, P., Chaturvedi, R., Kumar, S.: Advance approach for detection of DNS tunneling attack from network packets using deep learning algorithms. ADCAIJ: Adv. Distrib. Comput. Artif. Intell. J. **10**(3), 241–266 (2021). https://doi.org/10.14201/ADCAIJ2021103241266
12. Hong, D.: Codeblue Cybersecurity Conference, CODEBLUE (2014). https://codeblue.jp/2014/en/contents/speakers.html
13. T. H. News, Maldrone—First Ever Backdoor Malware for Drones, THN (2015). https://thehackernews.com/2015/01/MalDrone-backdoor-drone-malware.html
14. Almulhem, A.: Threat modeling of a multi-uav system. Transp. Res. Part A **142**, 290–295 (2020). https://doi.org/10.1016/j.tra.2020.11.004
15. Jares, G., Valasek, J.: Investigating malware-in-the-loop autopilot attack using falsification of sensor data. Int. Conf. Unmanned Aircraft Syst. (ICUAS) **2021**, 1268–1276 (2021). https://doi.org/10.1109/ICUAS51884.2021.9476717
16. Gorrepati, R., Guntur, S.: DroneMap: An IoT Network Security in Internet of Drones, pp. 251–268 (2021). https://doi.org/10.1007/978-3-030-63339-410
17. DeLaOsa: The promising yet vulnerable reality of unmanned aerial vehicles. ECN Electronic Component News **61**(2), 11–13 (2017)
18. Reed, T., Geis, J., Dietrich, S.: Skynet: a 3g-enabled mobile attack drone and stealth botmaster. In: Proceedings of the 5th USENIX Conference on Offensive Technologies, WOOT'11, USENIX Association, USA, p. 4 (2011). https://doi.org/10.5555/2028052.2028056
19. jgamblin: Leaked Mirai Source Code for Research/IOC Development Purposes (2016). https://github.com/jgamblin/Mirai-Source-Code
20. Tien, C.-W., Tsai, T.-T., Chen, I.-Y., Kuo, S.-Y.: UFO—hidden backdoor discovery and security verification in IoT device firmware. In: 2018 IEEE International Symposium on Software Reliability Engineering Workshops (IS-SREW), pp. 18–23 (2018). https://doi.org/10.1109/ISSREW.2018.00-37

21. Marais, B., Quertier, T., Chesneau, C.: Malware Analysis with Artificial Intelligence and a Particular Attention on Results Interpretability. Lecture Notes in Networks and Systems, 327 LNNS, pp. 43–55 (2022)
22. Iotti, E., Petrosino, G., Monica, S., Bergenti, F.: Two agent-oriented programming approaches checked against a coordination problem. In: Advances in Intelligent Systems and Computing, 1237 AISC, pp. 60–70 (2021)
23. Czyczyn-Egird, D., Wojszczyk, R.: The effectiveness of data mining techniques in the detection of DDoS attacks. Adv. Intell. Syst. Comput. **620**, 53–60 (2018)
24. Antonakakis, M., April, T., Bailey, M., Bernhard, M., Bursztein, E., Cochran, J., Durumeric, Z., Halderman, J.A., Invernizzi, L., Kallitsis, M., Kumar, D., Lever, C., Ma, Z., Mason, J., Menscher, D., Seaman, C., Sullivan, N., Thomas, K., Zhou, Y.: Understanding the mirai botnet. In: 26th USENIX Security Symposium (USENIX Security 17), USENIX Association, Vancouver, BC, pp. 1093–1110 (2017)
25. Hethcote, H.W.: The mathematics of infectious diseases. SIAM Rev. **42**, 599–653 (2000). https://doi.org/10.1137/S0036144500371907
26. Batista, F.K., del Rey, A.M., Queiruga-Dios, A.: A new individual-based model to simulate malware propagation in wireless sensor networks. Mathematics **8** (2020). https://doi.org/10.3390/math8030410

Author Index

J. M. Machado et al. (Eds.): DCAI 2022, LNNS 585, pp. 205–206, 2023.
https://doi.org/10.1007/978-3-031-23210-7

Author Index

J. M. Machado et al. (Eds.): DCAI 2022, LNNS 585, pp. 205–206, 2023.
https://doi.org/10.1007/978-3-031-23210-7